TABLE OF CONTENTS

Nathan Coppedge

Dimensional Time-Travel Toolkit

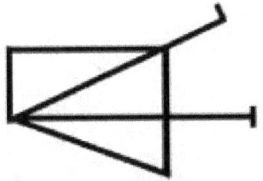

The Dimensional Encyclopedia
Vol. XV

Nathan Coppedge

THE DIMENSIONAL TIME TRAVEL TOOLKIT

Nathan Coppedge

Nathan Coppedge

DEDICATION

This book is devoted to the art of spell-casting, which begins to seem important.

Nathan Coppedge

Dimensional Time-Travel Toolkit

≈INTRODUCTION≈

Beginning on May 5th, 2013, Nathan Coppedge set out to write a book on time travel: the best of its kind. He borrowed his themes not only from science and semantics, but from a host of unique experiences which occurred during his developing life.

Several of these experiences seemed to involve genuine time travel, varying from 24 hours, to significant periods of years. Once Nathan observed his younger brother take a younger form than previously after seeming to time travel between different airplanes in mid-flight. In another case his mother told him that it was 2005, when the year was originally 2009. Incidentally, 2005 was close to the year he had intended to travel to. He then had an experience of synchronously re-living the majority of moments which had occurred in his life during that period, this time with knowledge to anticipate what would occur, including things that were said, and events that would happen. He felt a strange loyalty to the same sorts of timings and events that had already happened. But at the same time, he felt a small amount of willful control, even to change some details of conversations. In some cases, the recipients of his words expressed confusion, as if he had said two things at the same time, or as if the words had somehow been undone by a trick of magic, and re-written. In other cases, particularly with his father Michael and brother Brian, Nathan found that people were able to respond to different contents than had originally been expressed.

Over all the years that took place, including
some five years of double-history and two
time-warps, Nathan was developing a unique
set of mental tools, tools that were notable not
only for their contribution to temporal semantics and historical knowledge, but also for
their bona fide intellectual credo and sense of
jux d' prix. Nathan recognized that in all his
time traveling he was likely to un-write his
own history. He also recognized that there
was an important opportunity for personal
growth and genuine achievement. He recognized that other people could also be time-
travelers, even without knowing it. By adopting a dimensional philosophy, it became possible to recognize that conscious and unconscious forms of persons might exist---for
there was no way to excuse if someone was
confronted by a different time or place than
expected, and held to their original beliefs.
The simplest explanation was somewhere between an absence of time-travelers, and a thesis of total trickery, or chronological semantics. Because it seemed likely that if one person could time-travel, so too could others, or
at least that someone would predict travel
even if it was highly unlikely, and total trickery didn't seem to justify with having food
that was really food, and air that was really
air, etc., I often found myself falling into a semantic view of the chronology. Semantics is a
convenient view, because it shows that the
importance of a given set of laws or events is
no more than a degree of formal importance
ascribed to its properties. As I traveled, I
gained a total respect for the formal properties of reality. The paradox was that just as the
laws of physics might be semantic, so too

there is a reliance on semantics in the specific condition in which it takes place. Laws are, in the long scope of things, 'events': because immutability is really the countenance of something which takes place, if at all, through the actions of the things, persons, atoms, which compose the world that we observe.

I also came to have an appreciation of the open-endedness of the concept of God. Although in some ways agnosticism was a natural move for someone who had been raised Unitarian Universalist, in other ways I felt that my religious point of view was a direct function of my most radical experiences, in particular the subtleties of time travel. God----if we see him or her as real---necessarily can grant the powers that we observe. But, from a time-travelers perspective, different powers exist. The power to create life in one place is not the power to create that same life somewhere else. The power to time travel is sometimes like committing virtual suicide. Let me explain this: unless there's artificial continuity linking all worlds holistically to the same world, time-travel involves a schism between the experienced worlds of the past, and the experienced worlds of the present and future. The intermediate territory---perhaps inhabited by God, or unseen people, or some other form of life---is the territory in which travel takes place. Nothing at all about this, and at the same time, everything about this is semantical. Everything is, in some sense, either given or taken wholecloth by God. Time travel is like a blank check. Or, perhaps, one could see it as a radical communication barrier with aliens from another species. Or an ability to connect dots on a big, mostly white flow-

chart, which happens to be the moving-
picture of life, in its colorful blotty details.
Time travel, whatever the case, is a highly exceptional phenomenon.

Like many of the infamous criminal cases of
history, time-travel requires a motive. Like a
perpetual motion machine, it occurs when
there is sufficient momentum, and it often re-
turns to the same position. With rules like
these, time-travel becomes a science. Althoug,
without paying attention to these abstract
properties, its art form may easily be lost. It is
a subtle science. Part of my quest with time-
travel has been to record evidence that it ever
happened. But, unless I have an appearance of
un-naturally aging, or unless others share in
my experiences, it becomes highly difficult to
prove. Even if more than one person experienced the same event, it would remain on the
fringe of practical knowledge. Fortunately,
I'm not trying to prove to my mother that I am
a 30-year old version of her 5-year old son.
Like in the movies, the cases that are easiest to
prove are not always worth living. Remember,
that where time travel occurs, there is a form
of reasoning which justifies that it has actually taken place. Therefore, there is no way to
avoid the ignorance of those who would believe otherwise. They, too, are part of physics,
and to some extent their explanations also
hold validity. Where time travel participates
with relative space-time, its actions are palpa-
bly semantic, unless they involve externalized
technology. And the fact that technology tends
to be external limit's the effectiveness of technology in creating time travel. Probability and
continuity will always paint a different picture than time travel. It is easy to notice that

some people go missing, but hardly anyone usually thinks the explanation is time travel. Even if it were, it might be a fraction of the number of cases. So it can be taken for granted that time travel tends to be statistically insignificant. But, there is possible evidence. The person who is remains at a given time may have had strange experiences. But there is a tendency to explain it away as a form of madness or idiosyncratic thinking. Not everyone is ready to believe that those two pieces of evidence (someone missing, and someone with strange experiences) together mean time travel. Most people are about as willing to believe that time-travel exists as to believe in magic or demons. And they may even think magic or demons are more 'popular', even if time travel is real. However, in my view, time travel is not necessarily an evil thing to do. 'Diabolical' in a pop-culture sense, perhaps, but *not evil*. If tine travel is not evil, then it is hard to associate it with devils. The specific power to time travel may as easily belong to someone highly miscellaneous who is wielding a cusp of fortunate events.

Nathan Coppedge

"If we could travel in time as easily as we travel in space, we would become different beings…"

~~~Clifford A. Pickover: from *Time, a Traveler's Guide*

"It's about time to travel"
~~~The Robbers in the Woods

"Time doesn't pass as quickly as anyone thinks"
~~~Michael J. Coppedge

Nathan Coppedge

A SYMBOL OF TIME TRAVEL

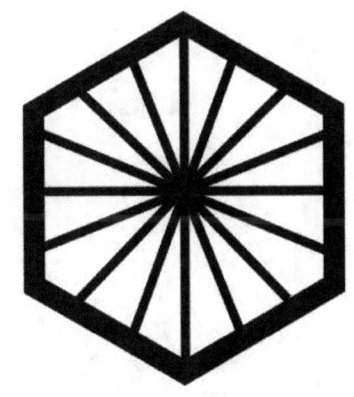

Nathan Coppedge

Keys:

[0] Conceptualism
[1] Semantics
[2] Induction
[3] Telekinesis
[4] Suggestion
[5] Power
[6] Art
[7] Alpha
[8] Mathematics
[9] Architecture
[10] Government
[11] Wisdom
[12] Time

Proverbs of Time-Travel

[1] "It's not gambling"
[2] "It's a reaction to ultimatums"
[3] "It involves desperate authority"
[4] "It contravenes history"
[5] "It involves motion in space"
[6] "It is an application"
[7] "Dispose of the riddles
        and learn to beg"
[8] "It is no safer than history"

Archetypes of Time-Travel

[A] DE PERPETUA
  (1) Infinity
  (2) Motion
  (3) Machine
  (4) Power

[B] MOT REGUM
  (1) Politics
  (2) God
  (3) Immortality
  (4) Logos

[C] RUET PENSIVE
  (1) Chronology
  (2) Mystery
  (3) Theatre
  (4) Magic

[D] VIA REGINA
  (1) Mythology
  (2) Nature
  (3) Death
  (4) Life

THE DIMENSIONAL TIME-TRAVEL TOOLKIT

A

Absurdity …plays a role in relation to time travel in two ways, and particularly in relation to absurd images: First, absurd images are a relatively strong starting point for mental time travel. They provide a basis for over-clocking exterior experiences, and thus gaining an advantage that can be used against time. Second, absurd images pose a difficulty to time travel, in that their physical existence generally complexifies the possibility of time travel. My evidence for this is that, absurdly or not, complex images indicate complex realities, and even more so than actually embodying 'time eddies', 'extra-dimensions', or 'motifs regii', they complexify the visual aspect of time-acquisition. So, some things to consider are that in some ways time travel is like a blind man dueling with the wind. In another way, one is combating the dubious magic of those who read images. In a third way, one is attempting to generate a very private visual language, which has priority. This could be said to follow from the same properties which make absurd images a good basis for time travel: the quality of seeming to resemble 'motifs regii', which can also be interpreted to mean a skry or 'tabulation' (amulet). However, I would warn that the magical perspective on time travel has little to do with the actual perspective required to 'encounter' (allow) them. The methods employed in this text require the mental rigor of a scientist.

ALTI-Matums: A form of time-travel occurrence. See under Matums.

ALTRI-Matums: A form of time-travel occurrence. See also under Matums.

(The) Anaxial Insight: This an insight that occurred rather late in my learning process. It
consists of the following smaller insights:

     1. What exists now is the essence of
time.
     2. Not only time exists, but also the
world beyond time.
     3. The failure of travel is therefore, the
failure of one-dimensional time.
     4. The more universal the time, the
more absolute the travel.

So, essentially, one-dimensional time can be
used as a guide to time-travel, using the relationship between time itself and the immortal,
changeless world. It then becomes important
to make contact with truth, in order to connect the absoluteness of one-dimensional time
with the more universal context of the time-
continuum. In occult terms, this may be called
'bullet travel'.

Dimensional Time-Travel Toolkit

Anomaly Rules

Based on my experiences, I have come to the following conclusions:

1. Coppedge's Anomalous Corollary: time travel does not tend to make big waves. For example, when you walk the street and ask someone what year it is, and they say it is thirty or more years before you thought it was, you might usually look to other explanations than time travel. On the other hand, if everyone is dressed differently, and there is no other explanation, then something happened. But clearly in that scenario the one making big waves is you---in that case, you are the one making the risk, and so it is less of a wave for other people. In either case, *relatively* there is no big wave, just a lot of little waves, in the case in which it works.

2. When extreme time-travel occurs, it seems to occur for a reason. You need a lot of reasons, or a lot of technicality, to actually travel. And the technicality or reasons are likely to be judged on their own terms. Traveling by reason is likely to be seen as mad. And traveling by technology is likely to be seen as an aberration---a physical exception or alternate dimension. So, once again, the properties of time-travel are not very exceptional. They are only exceptional by making exceptions.

3. However, it is not as if time travel does not have an effect. Whether or not your presence is taken-for-granted, you will be incorporated in memes, become part of traditions, etc. These things are seen as natural. However, you are making nature more like a time-

traveler. In that sense, you do make a big wave: just one: your sphere of influence. You might say: the big wave is the sphere of influence. Usually all that will remain as evidence are a few shreds of information, that are bound to be misinterpreted. Or you could make a mistake and end up in a mental hospital or deported.

4. In general, fate is on your side. Time travel (at least the mental kind,) occurs through incredible synchronicity of space and time. The more time and space synchronizes, the more likely it is that your basic needs will be provided for during the experience. Thus, it can be used as a desperate gamble for resources. However, psychologically, there is no guarantee that you will be in a good place. It is certainly a risk, and it goes without saying that it is not always a risk worth taking.

Aspective Differences: Small things may change between different times. Most typically, it is a thing easily noticed. A thing that differs in only one to three aspects, often aspects of the same types of properties. For one, shadowy, type of person it may be that a piece of art is missing from a display case. For another, behavior-oriented, person it may be that a large 'X' has been drawn across a poster. Signs like these are often forms of degeneration, and it is better to encounter them, as often happens, in secondary time-streams, which are later rejected in favor of a new avenue of travel. If you are clever, you can prefer a central time-stream where the art has been displayed, and where the large 'X' has not been drawn. But often a second step is to test

the illusion. You will need to make small sacrifices. You accept the types of signs that others would want to appear in their secondary
streams. You do not create them, you simply
allow them to appear. A strangely stylish graf-
fito appears, or the art for a certain gallery is
unexpectedly high class. This first shift beyond the secondary time stream is important
for gaining control of time worldliness. The
next stage beyond that is when categories begin to change. Perhaps the art gallery closes
down, or perhaps graffiti becomes less common. These kinds of shifts are natural, and go
hand-in-hand with a higher-quality existence, and greater control of time's modular-
ity, or if not, then at least a greater allowance
with those quasi-magical properties which
are closest to your desire. So: four stages have
been described so far:

    1. Secondary time ('degeneration')
    2. Primary time ('small improvements')
    3. Second step ('meaningless degenera-
tions')
    4. Categorical differences (art shop closes)

Perhaps I have not even yet perceived a stage
beyond these, but if there is one, it may involve contact with more universal properties,
greater object-reality, and hopefully greater
mental determination. One should consider
the types of ideas one might introduce to
make life more dynamic, and to serve as variables for whoever is doing the ultimate comp-
trolling of life. If one is suspicious of this person, however, it becomes difficult to make a
determination, and life becomes close to a
passive object. Release can be had through
moments like tasting ice cream in private, or

cuddling with a stuffed animal. These kinds of moments allow information to be collected psychically which can influence the ability to travel.

B

Bureaucratic Slight: If it is hypothetically possible to influence the use of information as
reflected back upon the person, such as by
such various factors as stealth, neglect, telepa-
thy, physics models, acceptable use, and temporary glitch windows, then it may also be
possible to make use of ambiguities in empirical information. For example:

1. The difference between 1-year periods and
the calendar year.
2. The difference between a one-day cycle
and the new day, or a one-year period and the
new year, such as ultimately by mistaking one
for the other.
3. The value of one particular date as a holi-
day or occasion for time-travel versus another
day.
4. Differences in the time system being used,
particularly on subjects such as the beginning
or end of a calendar, the use of disuse of a
system.

Generally, the method involves out-thinking
the time-system. Therefore, the primary need
is to suspend an indefinite ability, rather than
to produce some effect which is not detectible.
If a permanent window can be secured, the
method is much more serviceable. For example, by merit, authority, or preferentialism.

Burning the Past to Feed the Future: One of the insights that connects time-travel to immortality is the ability to operate crunch work on the past, producing an efficiency in which the future has not arrived yet, although it seemed that it would. This is particularly true with the sense of artificial reality or action-points systems, in which for example, speaking or moving between rooms can seem more time-consuming than they should be. Performing the crunch work, which is like shrinking the past, the future often has not yet arrived, and the effect is an efficiency. This is called 'burning the past to feed the future.' It can be used as a general principle of semantics about time-travel.

C

Cape Scott Case: See Also Appendix II.

Originally published erroneously in Peterson, Lester Ray. *The Cape Scott Story.* Mitchell Press, 1974. *Photo originally observed by Neil Bhatt as shown at Quora.com.*

Causality and Imitation: Time sings one song, and we sing another. This principle is deep enough to describe some of the properties of time travel. Essentially, we have a choice. But to be clear, we can always sing our own song. And there is always a causal determination of the song. But sometimes time is complex: dimensional. If time can travel, then so can we. Thus, sometimes the song of time is bifurcated into multiple choices, which are still part of the same song. In other cases we are simply being by-products of the chain of events which led to the song. So, we would do better to imitate time than to come up with our own song, if we want to time travel.

Chrono-Logic: Depending on how you travel, traveling into the past or the future can have different effects. This is perhaps best understood in terms of chrono-logic, or that is, logic which is determined in terms of age and the amount of time that has passed. The amount of time that has passed can be viewed semantically or symbolically, so that, given the uncertain nature of travel, it is always uncertain how much time has passed. However, one can guess, and then make suppositions based on what one believes has happened, or what *could* happen. Thus, chrono-logic. Here are the basic categories: (1) Atemporal location (the present): (a) gain wisdom, or (b) spend time.
(2) Future Location: (a) gain experience, or (b) save time.
(3) Past Location: (a) gain authority, or (b) gain time.
The nuances are somewhat subtle between gaining time and spending time. Usually they are seen as the same thing. However, there is

actually a subtle difference. Gaining time is sometimes like saving time, if it has authority. That is basically the difference. Thus, travel to the future is sometimes valuable for traveling to the past (since gaining experience can be worth authority). Similarly, atemporality can be worth time-travel points if it is the wisdom variety. Saving time can also be used for spending time, and spending time can sometimes be used for saving time, if there is a method of bargaining.

Conformity and Time Travel: In an absolute degree, when conformity is proven to be an unthinking process, it also has the effect of reducing creativity. Assuming that conservatism is perfect, it can claim three out of four categories of time mastery. But there is a lingering question about perfection, and with the fourth category, which involves the willingness to parse reality and translate dimensions. Perhaps one in four categories (the imperfect, non-conformist case) is not enough to begin study of morphology or pan-dimension. But using only one category of continuity is the conservative view of creativity. It assumes that time travel is not casual, and it also assumes that it doesn't involve much on-the-fly thinking. The conservative view is the contractual view---and accepting that it is creative overcomes the problem of beginning with only one category. So here is a list of pragma:

1. One creative category for institutional time travel.
2. Three creative categories for adaptation, four categories for learning.
3. Two categories standard.
[4. Fourth category is any available exception

to this.]

So you can begin with exceptions (the simple case), or you can begin with other people's knowledge (the institutional case). Or, you can be tough and try the knowledge / omniscience case. But typically some adaptation is involved, which implies a three-category though process. In general, a mixture is the best bet. All four types are strategies which must be used whenever they work, if you're trying to travel. See also The Fallacy of Non-Conformity.

Corporate Dependence: In the technical jargon of time-travel, there are several types of 'corporate dependence'. Corporate dependence is a reliance on a 'corporation' or association between two bodies. The body that one is dependent on can be called a corporation also. So, for example, one can have a corporate dependence on one's own material body for time travel, if the body is one's access to time. According to spiritual theories, Time may be a god, in which case one's access to time is through psychology or the spirit-world. There may also be corporate dependences upon time directly, which occur during travel. It is this final sense that I wish to emphasize, as it is more emphatically related to the nature of time-travel. Corporate dependence in the sense of a reliance on time means using time-travel techniques in the following ways:

1. Between matter and matter vis. Time,
2. Between matter and Time in terms of space-time-energy, and

3. Between Time and Time, in terms of matter.

These three approaches can be useful in the abstract, although putting them into effective use often involves a second level of perfected technique.

Counting to Two: A basic method for conceptualizing thought-travel is to consider counting from '1' to '2'. By the time the second digit occurs, one can consider that it might be ambiguous, in those terms, where the first digit occurred. If that is the case, the first digit might be located far back in time, in the recent past, or even in the future (if you had traveled backwards already while counting). However, consider this additional information: the longest time that transpires follows the most direct path through time, as indicated by cause-and-effect. This is true whether it occurs subjectively, objectively, or both. It is true regardless of judgment. If that is the case, the most obvious interpretation of where the first digit occurred moving backwards from the '2' is the '1' that is located furthest away! So, if it is ambiguous when the '1' occurred, then one may as well gain more seconds than were lost! This is particularly true if the future is less ambiguous than the past, e.g. because at some point you have to choose when to say '2'. Following this is a diagram to illustrate the selection. Notice that time is not very different for earlier or later '2's, but only for the '2' that occurs at the exact moment of being the result of the chain of cause and effect! This does something to explain time, and it also does something to explain time-travel.

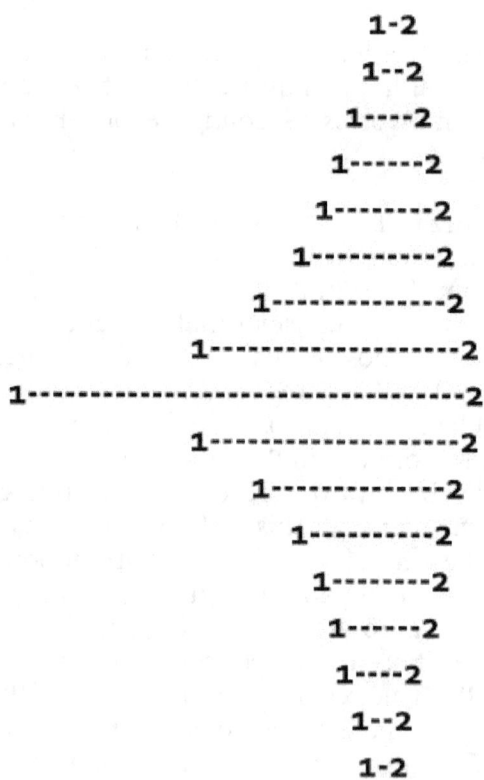

```
            1-2
           1--2
          1----2
         1------2
        1-------2
       1---------2
      1------------2
     1------------------2
   1-----------------------------2
     1------------------2
      1------------2
       1---------2
        1-------2
         1------2
          1----2
           1--2
            1-2
```

[The] **Criteria Problem and Inescapable Difficulties**: A series of grandfather experiments were conducted by physicists, e.g. with atomic particles to confirm the David Deutsch hypothesis (using a quantum explanation for time travel), or Seth Lloyd's alternate model. Both hypotheses have been confirmed, supporting Deutsch's hypothesis that the probability of destroying a grandfather is equivalent to the probability of becoming a murderer of that grandfather and a time-traveler, and under Lloyd's theory, that travelers may favor a world similar to the one they formerly lived in. According to my thesis of the criteria problem, those who time travel to a given location tend to resemble those who live near that dimension of reality. In fact, there is a tendency towards 'radical explanability' that is more than just coincidence. This is even, or especially, the case for time-travelers, in part because travelers are more likely to be explained away as tourists or oddballs, and there is sometimes a willingness to consider people with unusual explanations 'mad men,' thus eliminating the efficacy of some of the few (probably rare) conditions of confirmation. On a quantum level, following the evidence confirmed by physicists, convenience is not only a tendency, it is a rule. In other words, there will be no case where the energy available to measure aberrations exceeds the energy used to create those aberrations. As soon as aberrations are marked, they themselves serve as explanations. The inescapable difficulty is a second property, which states not that people are similar who are near a particular time location, but rather that travel grows increasingly rare with distance. The obvious corollary here is that on a macroscale, quantum properties are also being observed. So, likewise, and perhaps to a more radical degree, there is a sort of 'quantum conformity,' which permits time aberrations only when they are explainable. Another rule may be that when explanations are not found, instead no explanation is available, and these aberrations tend to be able to be explained away (this can be called differential unexplainability). Again, there is a limit on explanation relative to the strength of the aberration.

D

The Darwinistic Problem: There is a general problem that is fairly unique to time travel, namely, what if within some system of mind or matter, there is someone who is thoroughly advanced in every possible way---except he doesn't now how to time travel? This can be thought about Darwin, if his theory of progressive advancement applied to himself as well. One can imagine that someone only advances by moving forwards through time within Darwinism. The most typical sense of this paradox is that evolution might not evolve if it had to evolve backwards through time---in other words, evolution is also anti-evolution, unless it incorporates time-travel. So the choice is to accept instead, that, as far as temporal beings go, Darwin is better off than average, or to make some bizarre conclusion about how Darwin, for absurd reasons, doesn't possess the ability that others do. In some ways, the absurdist angle is the kind of things that might be reserved for fictions rather than facts. So, the following is concluded: (A) Darwinism expresses a kind of fact, (B) Dimensionally, time-travel expresses the relativity of time, in other words, it is important for measuring and systematizing time, (C) Consequently, time travel may incorporate Darwinism through exceptions in time, but (D) If we encounter Darwin, we have to decide whether we are also encountering his theories vis. Time and evolution, or if his theories are merely objects which can be handled as though they do not make changes across time. In other words, there is a question, finally, of entity-dependence, which suggests first of all, that most thought constructs are radically physical, but secondly that there may be networks of exceptions and associations which DO in fact have implications (or, as the case may be, *some* implications) across numerous time schemes.

Desirable Bargains: Making bargains with yourself, or personal sacrifices, is one way to add up to a time-travel situation.

Having Lots of Free Time: This may seem like an evil bargain if you don't know what to do. Instead of feeling horrible about yourself, you can decide that 'there's always something left to do' and try time travel.

Saving Productivity for Later: Having a future productivity goal can motivate time travel, because one can decide that one is 'doing some research' or that the bulk of one's accomplishments will not be swept away in the time-stream. So, productivity is generally lucky, as long as it is waiting for you.

Having Lots of Things that Handle Themselves: If your life is 'fully automated' then there is some freedom to travel to the past or the future of the current life. It may help to have a file where your passwords are saved, so that you don't have to remember both your future, past, and present versions of your passwords. Not that time-travel happens a lot, but it might happen (and it has happened to me).

Determining that Life is One Super Big Ongoing Project: Remaining forwards-minded is highly useful, and so is making a long-term commitment to your own life-process. It may help to know that a few sacrifices may be okay. There may even be ways of compensating for some minor, or even major, disasters. Remain confident that something remains of your life, and the time-streams will carry you to your next destination, however distant, obscure, or valuable it is.

Dirty Secrets to Time Travel: Some secrets are not pleasant to learn, but they are nonetheless highly useful.

1. If you time travel, you may be the same person in a different place. This has the effect of being unfair for those you meet who haven't traveled, because they may know you previously as 'God's perfect pitch': something that was just loopy and playing along before the real character arrived. Conversely, the people you meet when you time travel are more likely to have traveled themselves than you are, so in this sense, they don't see you as 'God's perfect pitch'.

2. Many people are anachronistic to fault before they time travel. This sometimes doesn't mean they live in the past. Sometimes it means they are hopelessly futuristic. This may mean that time travel always involves travel to the future, although often in a different dimension. However, relative to other individuals who do not make a choice to travel, or who do not accomplish anything over the years, the futuristic person may appear like a figure

from the past.

3. In the case of bureaucratic slight, it may help to be ugly and ambiguous in appearance, so as to be 'conveniently lost' from reality. In this sense, the skill of time travel involves being in touch with the continuum between one reality and other realities nearby, involving memes or tropes, etc.

4. One thing that can encourage time travel is if someone has a grudge against you. You can use a grudge as a reference point for deciding to leave the world as you know it, or you can even use the motive to recover from a slight as a guiding point for finding someone who may not even be the original person you intended to meet. Time travel thrives on this kind of serendipity, although it doesn't thrive on attachment without considerable power.

5. Stealth is another overlooked factor. Public attention can inhibit the ability to time travel, because of the assumptions other people have of what you are capable of. There is a learning curve with fame that may require more cleverness. Conversely, public attention may sometimes generate the right type of energy.

Divine Manifestation: Ponder: 'They thought history traveled in a line'!

Now that you see history manifesting in a curved line, you can choose to manifest at any time to the people of the other dimension. If life is as plain as mathematics, though, there are not likely to be very interesting people there. But if life is philosophy, or politics, or psychology, then you are likely to know what I'm talking about. There is some *substance* some background there----and the less straight-jacketed you are about your idea of *history*, the more likely you are to appear like a *God*. You may never *know* these people, but rest assured, *power* usually conveys a *good* vibe as long as it's mostly entertainment. And *entertainment* is encompassed by the idea of media---which is even scientific media! So rest assured that you will make a good impression if you manage to manifest in an alternate dimension! Life may depend on it! Charm the socks off of them! And if you don't know what happened, you can always go back to philosophy or anthropology or linguistics! Feel free!

Divine Method of Time Travel
Type 1: Reversion Using Spots of Time: This method requires a high degree of technicalism, more than any of the categorically numbered methods. As usual for this type of advanced level, it may be viable to consider that time travel is a gift rather than a talent. However, I will continue the instruction under the option that it's possible to teach *something*: If you know what time looks like metaphysically, e.g. like lemon-stained burns on a bulb of glass, then it may be possible to study the immortality of the burns, and then travel to other temporal locations of the burns. However, what makes it difficult is that the burns have no physical location, so theoretically this requires some degree of omniscience. If the bulb is a crack vial, the advice is not to take the crack, because that's where the time comes from, not where it's going to. We are

looking for immaterial stimulants, not material ones. We want the highest degree of benevolence on subjects of recovery and efficiency, which can only be had with an unaided mind. (According to relativity, just because someone else looks divine taking a drug doesn't mean you will too). According to this model, all ideal experiences occur outside. All ideal experiences occur on the surface of the world. And theoretically the crack addict doesn't know this. Nor, given the difference between the addicted mind and the unaided mind, *could* the addict know. However, at earlier times in history, other objects have also had a similar role, suggesting a kind of arbitrariness in the choice. For example, glass spheres of the earth were at one time rare, and were more easily mentally manipulated before the parasitic addicts came into play. Before that similar effects were had with water or sunlight, although not always knowingly. For example, instead of time travel, it might seem like a change of landscape. That's how wealthy in time these beings were. The most likely explanation of the crack addict is that his or her sense of time is below average, and he or she needs to take the drug just to feel normal about the pace of life. E.g. an only-when-its-justified-in-any-way stimulant theory. Versus the it-signifies-death-no-way contrary theory (it should not be necessary to warn you that crack might be bad for ambergris, or if it is an ambergris, it's much better for god than men). Perhaps it is only when God uses crack that it becomes an agent of time travel for those who feel like mortals. Those who feel like immortals, but without strategy clearly lose out for good reason. Similarly, the crack addict justifies his or her own experience, yet without any expectation of reward for this behavior (except, initial innocent theories). Essentially, the crack addict is giving, but the giving has nothing to do with crack. It is both cruel and efficient, and almost no one benefits. But the benefits are extreme for those few who learn how to make use of spots of time (without using drugs).

Doing the Wave: This may not be a visible wave, but instead a pattern of energy which spreads over an area associatively. The uses of this are fairly obscure, and the frequent invisibility of the effect may well be due to the limitation of human perception, namely that the eyes, which are normally our only apparatus for sensing this phenomena, frequently apply automatic corrections to avoid disturbances which originate internally rather than externally. I have found 'the wave' useful on one occasion, which I mention in the section in this book titled Time Travel as a Survival Technique. Essentially, memory was an effective wave pattern which I could apply in such a way that I did not need to remember the whole world to re-orient and re-locate myself within my context when I was about to die. I used the momentum from my fall from a building to re-ignite my existence, and used memory as a keycode to orienting myself. With that much described, I admit the instance must be rare. But it seems to prove the existence of the wave.

The Doppleganger Experiment: Once when I was in elementary school, I remember a so-called friend taking me to my future high

44

school. And I remember the high school looking as it did not look for a number of years
after that in the original time frame. When we
visited the school, I had the creepy sensation
that I would meet my future self. In that moment, I assumed that we (my present and future self) would somehow annihilate each
other, or I would go insane. Incidentally, I did
eventually go insane, although it wasn't until
after I passed through the time frame of the
self I thought I might annihilate with. But the
point is, when I learned to time travel later on,
I thought about that earlier moment, and discovered the probable rule of what it means to
meet your double from the past or future. The
rule is: (1) If you're traveling to the past, first,
your time-travel self tries to remember if it
decided to time travel sometime in the past. If
it did, it identifies the past self with the self
that decided to time travel. The communication becomes about time travel. If the past self
does not identify with time travel, then the
current self determines that the past self is
someone else. If one is traveling to the future,
the future self is necessarily less real than the
present self, but this poses problems for the
reasoning of the current self. The current self
tries to warn the future self to play catch-up
time and develop an educational experience
about what was learned from the new time-
travel version of the self. But if the future self
does not learn and play catch-up, then the
past self suffers from extreme shock, developing insanity. The contact with the past self
may even destroy the future self, since it is
now considered expendable unless it is a different person. (2) If you meet a time-traveler
from the future, the effort is to learn everything about your future. You are likely to

identify with the time traveler, but also to act confused, and then forget that the event even happened---because it was so strange. You develop a strange forgetfulness that only dis-appears when you or someone else finds proof that the time travel took place, often later, af-ter having already repeated the experience from the other angle. If you meet a time-traveler from the past, you are less likely to be confused. It might be an experience you have already had from the other angle. If not, then you can deduce that it is a time travel experi-ence that you're going to have, whether you want to or not. Therefore, meeting a genuine double from the past is a definite sign of being in the hands of fate. However, this period of being in the hands of fate is likely to last only until you yourself learn to time-travel.

E

Encloistered: A means of containing localized concepts, and also a strategy for projecting into new events. Essentially, thoughts and subtleties of physics are parsable in clusters. A speed function might resolve the significance of the cluster, but mastery of the cluster is what is entailed by local mastery. Mastery of the cluster allows some windows to open about 'purely local' events, but not broader events. Thus, stretching the local function results in extended functions in space-time. Although the potential of clustering is exceptional, and limited by the range of the function, particularly its ability to apply to exceptional details of greater ranges (represented by the individual's historical or personal knowledge), the primary role of clustering is essentially limiting. In other words, clusters are essentially summarized by the range of knowledge of the individual.

Ends
It is important to distinguish that 'ends' are definitive of the past. This is a distinction from 'goals', 'attempts', and 'keys'---which tend to exist in the future---and objects and fates, which tend to exist in the present; The distinction is important because of the role of location in strategic planning. Immortality aside, past objects called ends are important for strategic allocation of temporal real estate, initially the most essential aspect for the mastery of time in any larger sense than physical

47

immortality. The object of the future con-
tained in the present end is a key to wisdom,
luck, and intelligence, sometimes called psychic powers. The conjunction of the future
and past objects is an access also to psychic
mastery, better known as authentic dynamic
intelligence. Although, there are ways to exercise this without traveling great distances, or
at least not obvious distances. This is part of
the theory of experience, incorporating gov-
ernmental concepts of time, and seems
broadly economical, hence putatively bureaucratic or subject to mathematics.

Entering the Tunnel: Politicians have an idea
called 'entering the tunnel', that is not dis-
similar to the idea of a deer caught in head-
lights; Sometimes, however, entire scenes,
events, and moments in history are caught in
a whirlpool---they have entered the tunnel;
The minimum appreciation of this is that everything cheap is put on high liquidity; This
means that everyone above you has significant
opportunities to profit, and you can only win
if you're sacrificing, or if you have authority;
Consider, however, that it is possible to leave
the tunnel (Not by dying, but rather by an act
of significant understanding, what is called
'second guessing');

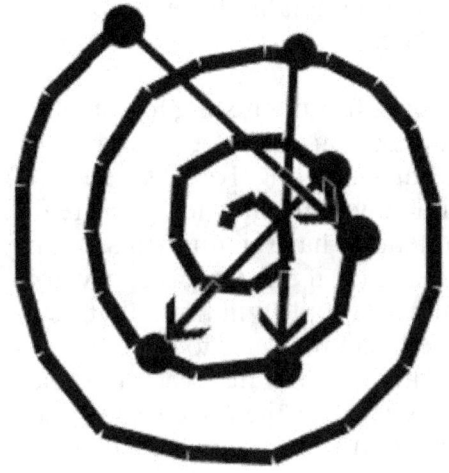

"The Tunnel"

The sooner you leave the tunnel, the less people gain from you (this is all very subtle: ordinarily there is nothing to gain from the tunnel, since it is 'only an idea', but this is an encounter with fear and realism, one of which is an enemy, and the other of which has limitations); So, the tunnel has a sense of economics; As soon as you understand the tunnel, they're on your watch; Thus, the longer you wait I the tunnel, the better bargain you can pull off if you're clever; But if you're not clever, it's assumed that 'the tunnel ate you'; Thus, what emerges from the existence of tunnels is people who have a small number of exceptional advantages; Time travel becomes a prospective high-level currency which is especially valuable, because it tends to formalize metaphysical trickery; But since anyone who doesn't play clever is lost to the tunnel, formalized trickery is actually a good option within reason, particularly because those

that make the rules tend to have more ex-
travagant taste::

Entic Exception: It is interesting to note that
an individual person might be the only exception to not time-traveling; For it is true that
time travel involves multiple temporal realms,
realms that tend to have significant similarity;
What, for instance, if one travels backwards
by ten minutes?; People will think that the
person is identical, not refer to the later for
proof of the missing minutes; The difference
will go unobserved, as the traveler navigates
the similarity to his former place, which appears in high-contrast; I argue that this only
takes place by taking the role of the former
self, because this is the form of the body that
has the greatest likelihood to exist at the same
time; If the traveler's destination is the future,
observers will assume that time transpired,
and ignore differences which appeared between the previous internal experience and
that of his would-be self; Even God seems to
justify or de-emphasize the presence of a
body, depending on whether someone has
traveled; For example, traveling into the future seems to eliminate the future body (e.g. if
one ceases to exist in all neighboring worlds,
when the origin of those worlds is proven to
be the local past); In this way, perhaps it becomes a 'traveling world'; Similarly, the present of the traveler is likely to be one without
a second self; The traveler may even feel that
it is easier to identify with others after traveling::

Dimensional Time-Travel Toolkit

Excuses: According to the media, the reason to travel often involves criminal motives or some form of heroic derring-do. In reality, time travel exists on multiple levels. The excuses for one part may be different from those of another. Often, in real life, time travel is justified from multiple points of view. Here are some examples. But you will notice, none of them are actually heroic.

Honest motives:
1. Curiosity / Exploration.
2. Self-Erasure.
3. Immortality.
4. To re-live one's life.

Criminal motives:
1. To escape one's past.
2. Greed.
3. Murderous intentions.

Excuses for assisting someone:
1. One owns the time stream.
2. An intelligence test.

God's excuse
1. It's still time, no matter what.

Exegency Factor: The rough pre-ordainment of events by matters of birth, priority, or circumstance. In a more complex fashion, it also considers other types of factorization, any of which become the default context or 'loose module' for time travel. Examples of these include education, stories, trickery, and magical authority. Exegency, or exegesis-ness is a term that means the prior on-coming of some factor, such as a system, agenda, or variable.

F

Fallacy of Non-Conformity: Non-conformists sometimes feel like mobile citizens, but this is not always the case; They may feel isolated from income or classicism. This conforms to the trend of being locked out of some of the more significant institutions and constructs which might promote time travel. Perhaps this prohibited quality is what contributes to the non-conformists' supposed attraction to illicit drugs (a form of disaffection). Certainly drugs seem to provide a deconstructed rather than affordable path to time travel. If there is a good quality to drugs, it appears to be the level of comfort someone attains with them (something that is by no means automatic) and the reliance on solutions made by the drug in place of the person (passivity or despondency). This involves considerable trust that there is a mesh between the specific advantage of the person, and the associations latent in the substance (i.e. 'appropriate deviation' may be 'ironic'; 'ironic' as a positive looks like luck, and luck looks inherently desperate). Perhaps that is why drug use is frequently called 'abandonment'. It opens up a life to considerable conceptual difficulties, which have the effect of artificial zing what may otherwise---if not achieve by drugs---be a very authentic, although challenging, time travel experience. In this respect, I tend to take a Zen and literary ambergris approach, which has more to do with subtle effects and grand mastery then it has to do with evoking rushing feelings or 'trippy' transformations. If someone travels effectively by illegal drugs, it is likely to be a product of paths laid out by subtler thinkers, or an application of wasted talent. One may ask, for example, how such an experience could ever be 'subtle' to say nothing of 'recoverable'. In an age of relativity, there ought to be a deterministic law against using drugs for time-travel. The overall effect of their use is likely to be destructive.

Five Types of Time-Travel: This arrangement clarifies some of the connected functions of time-travel. See the following diagram:

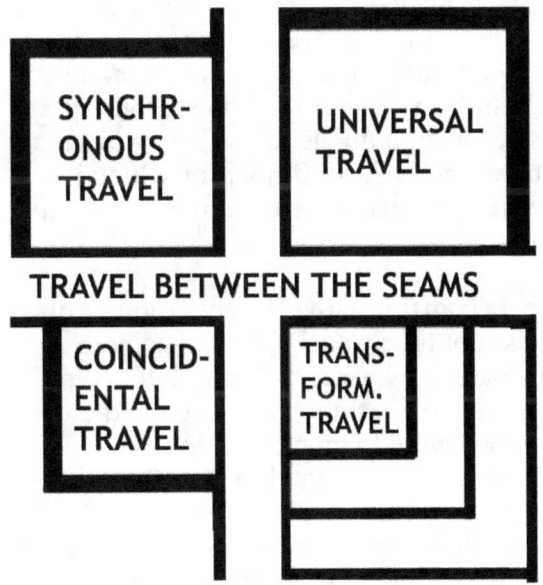

Type 1: Universal Travel. This is also called entity-travel. It involves ordinary choices about where one is located, but can also involve volitional choices about the composition of the manifest world. For example, the appearance of a person on a particular morning, such as what clothes they wear, and how tidy they are, can affect the appearance of other

people around them.

Type 2: Synchronous Travel: This is the type
that involves connections between different
dimensions. One may have to guess how other
dimensions appear, or travel to a past or future of the current dimension, which is also
considered dimensional travel. Essentially, objects are connected through similarity, so that
there may be more object coherence than time
coherence.

Type 3: Coincidental Travel: As an extension
of synchronous travel or universal travel,
when one thing, or coincidence, is modified,
this results in different manifestations also.
This becomes valuable for traveling distances
that are utterly synchronous, or which involve remembering the condition of the past
or the future.

Type 4: Transformative Travel: Another dimension of time travel is the capacity to become older or younger, change hair-color,
etc. Ideally these things would be under control if one were to time travel for long periods.
Otherwise, one can get 'lost in the seams'.

Type 5: However, the seams themselves offer
another possible mode of travel, which is to
allow time to change, while oneself remaining
still. Or to become a former self and re-live
the same life. This form of time-travel is relatively advanced, but some may consider it
easiest for travel without wishing, because it
involves the real connections of existence,
such as one's own body, or the disconnection
of time.

The Floating Bottom-Line: I try to see things now and then from a social science point of view. According to this view, time travel reduces to whichever key concepts capture its problematic functions in the context of statistical calculus. An analogy can be drawn to calculus in that the derivative is partly a function of a line segment (like time), and that this derivative is essentially floating in space. Now, as the line segment gains complexity and density according to a hypothetical 'paradise logic' (e.g. dimensions expand by functions and entities), The curvature of the line-segment becomes increasingly comparatively simplistic. One can imagine a sculptural metal curve enlarged and simplified into a truss bridge, which then becomes a suspension bridge---maybe not simple math, but there is a simple grace to its functionality. What I am reaching is, the node of time is increasingly ersatz, however complex and perfect it is, an-however it's context is increasingly one of equilibrium. This is an interesting figure, mathematically and otherwise, and combines well with the paradise logic. Essentially, the ersatz value of life-being-lived, regardless of exceptions, is provided for to the extent that it is both perfect and complex. What this means is that although a life must be complex to get complex results, it also must be perfect to get perfect results. And although the interior reality does not always reflect well, it often reflects functionally in the outer world. Now let us apply this to time travel: things are not necessarily about the right reality or the right technique, so much as 'the right reality technique'. Notice that the combination may have less social-science evaluative value than the sum of the parts. Therefore, it is efficient.

Therefore, it is possible to see that there is a broad range of functions which benefit from combination. Combinations between reality and techniques, perfection and complexity, and functions and entities serve the calculus of time travel. However: here is the shortcut: everyone is already, relatively, living it. Solving the equation is no more difficult than extending a line with an approximation, or reading the calculator or instruction book which has already been provided! Clearly, calculations in this sense require values which are already present, an advantage which pre-formulates some opportunities while leaving others powerfully open.

Foreshortening or Trapdoor Problem: A potential problem arises in the joining of one moment with the next: without complete devotion to the continuity of time, the result is to lose efficiency, however, complete devotion to one instant creates imbalance. Temporary advantages that emerge from strategic efficiency are not always guarantors of progress against time (although certainly they can be). One shouldn't consider the struggle against time to be some sort of sacrificial bargaining system, or at least not without the expectation of gain. Instead, there are several more generalistic strategies, strategies which are not directly related to the problem of patience and devotion to continuity. One of these is simply to promote personal health and well-being, to be conscientious about time, and to approach life with a leisurely seriousness, which combines these aspects. Some would say, however, that this is a strategy without a technique. It lacks a specific method of time-travel, and thus--- at least in the context of time-travel---it may also lack a method of ultimately competing against time. Although the most patient methods of competing with time are described in a later book called *The Dimensional Immortality Toolkit,* in the context of time-travel there are a number of methods for competing strategically against the foreshortening or trapdoor problem. To believe the idealist position, there could be ways in which complexity is actively sought out specifically within trapdoors. Perhaps falling through a trapdoor is a way of failing a test (such as a game) which has been proposed by the makers of temporal reality. Thus, an obvious solution emerges for improvements of value, in addition to improvements of health and efficiency. Clearly, however, such values which do emerge cannot stomach very much risk, unless this risk is specifically a recourse to the desired method of travel, health, efficiency, etc. I can imagine a case where this is so, for example, if there is total devotion to time travel, and value is greatly improved, and then efficient value is devoted to gradual changes in temporal security, for example. These are typically cases which do not involve much anxiety. There is a principle with the idealist position to question reality where it does not matter, but to follow conventions in all matters of risk. Another principle besides the seeking of complexity within the trapdoor is to seek continuity in all places where trapdoors are not encountered. If additional time can be generated when trapdoors are not present, then it seems possible to survive trapdoors with an acceptable loss. Although, it is also possible that due to health or accident, trapdoors are not the greatest danger at all. So this advice for generating time is subtle enough to only make

sense when health and safety are already secure. A still further method is to correlate
trapdoors with allotments of time, but this
runs the risk of requiring constant mental
oversight. It is a so-called 'ambergris method'
because it begins by assuming a kind of divinity status, and extracts minimal subsistence
from that.

Forms of Repetition:

Arbitrary repetition: Can be circumvented by
arbitrary circumvention.

Random Repetition: Can be circumvented by
re-interpretation.

Ambiguous repetition: Can be circumvented
by adopting a point-of-view.

Perspective repetition: Can be circumvented
through a meaningful interpretation, such as
mythology or a plot drama.

Forward Travel (Archetypal Methods): Time
traveling forwards may not require quite as
much patience as traveling backwards, but
patience is still required. One way to seek
loopholes forward through time is simply to
have fun, following the axiom that 'time flies
when you're having fun.' However, to get
consistent results, one may have to go beyond
mere patience or fun-having. Time traveling
many months is a rarity. Usually, the only
way to travel forwards for months is by time
traveling backwards, and then living the ordinary way for the rest of the time, or else
blacking out, and having other people manage affairs. Those two views are not exactly

forwards time-travel, so understandably, one might look to other methods. Nonetheless, that much can be kept in mind. If time-traveling backwards requires losing the present reality and exchanging it for an earlier one, time traveling forwards is more difficult, because it involves accomplishing everything that a future self would accomplish, in one instant, or else accepting that one is less accomplished than one would normally be. That is a difficult compromise, one primary reason why forwards time travel is not always desirable. For most purposes, moving forwards is the way to 'experience' time, whereas moving backwards is the way to 'travel.' However, there are prospective methods, such as the following, borrowed from an occult tradition:

1. Weaving Invisible Threads: Making and unmaking time itself.

2. Lifting the Hem of Time: Traveling underneath surveillance of time.

3. Walking Down Age's Stairs: Symbolic passageway of time and place. This is similar to Gulliver's Seven-League Boots, only concerns time.

All of these three methods are direly distinguished from auto-aging, which is a magical effect which doesn't actually spend time.

Experience with Traveling Forwards Through Time:
<"I'm Jesus" he said
>"I know you're Jesus because you spoke with the voices in my head."
<"Anyway, I'm more like an avatar type, if you've heard of those. People have things called cellphones. Do you know about those?"
> "Yeah, they have them in 1999 but they're pretty clunky (in 1999)."
< "With no further ado, I will show you 'you'!"
> "Not in person! Well, show me proof!"
< "You will take the cover off of your ventilator and rotate the fanblade counter-clockwise."
> "So detailed! But we won't see. I mean, we won't see?"
< "You won't see until you live in the future permanently! But you don't nee to see if you know!"
> "I still want to see! With my ears, as I usually do!"
< "Very well, we happen to be in the right neighborhood. We're going up to the chimbley so to speak. Hold on!"
> "Here?"
< "Anywhere!"
We appeared on the roof of a building I thought I should be unfamiliar with.
< "Hit your hand on the thing with the peaked roof...."
> "It's called a chimney."
< "But, make sure you make your unique sound."
> "I did, I know."
< "Do you think you can feel that he heard?"
> "He heard." I said.
< "Then it is time to go back…"

Dimensional Time-Travel Toolkit

Foundations of Synergistic Time:

Cohegency Thesis

If time traveling backwards creates funda-
mentally different experiences, then why does
the same artist create the same painting twice
(as I have observed)?; One solution is that
both cases involve the same fundamental experiences; The individual may be responding
psychologically to macro-factors in the world;
Another interpretation is that the context is
artificially constructed from templates; In this
book I will take both theses seriously, as it is
possible that variations of both general conditions exist as possibilities (vis. Modal realism)::

Four Methods of Time Travel
(the hardest and easiest methods)

*Time travel by marking a word
(*the most difficult method*)

This includes the following sub-
methods:

1. Choose a word such as 'old' or 'young'
or 'time';
2. Mark the word by thinking about what
it means;
3. Think about the opposite word;
4. Travel to the opposite mark;

*Time travel by negligence
(*second easiest method*)

This includes the following sub-methods:

1. Follow the path of life
2. Notice what conforms to the path
3. Identify conformity with time con
formity
4. Unconform

*Time travel by not counting
(*the second most difficult method*)

This includes the following sub-
methods:
1. Imagine counting numbers;
2. Imagine counting backwards;
3. Imagine counting into earlier
numbers;
4. Imagine that if infinite numbers are
counted, you will not lose infinite
time;

*Time travel by chronology
(*The easiest method*)

This includes the following sub-
methods:

1. Be born
2. Don't lose consciousness
3. Observe the world
4. Age

Fractionistic Origin of Time Travel: It is sometimes suspected that because of entropy,
things get worse with additional time travel
experiences. However, I find this isn't inherently true. For example, there is an illusion
that develops about structure and architecture, that these things are best understood in

their origin, and without dynamics. Usually that is untrue. Anyone who benefits in real terms by exploration (be it informational, physical, or psychic) will also benefit by time travel. Mental exploration by time travel also serves to benefit knowledge, as it is a more dynamic and living view of experience. The danger here is formalism which has been imposed to artificially simulate time travel, as for temporary benefit, or to serve a nefarious purpose. Some level of originality, such as the concern for 'art' and 'alpha' (see Keys in the Introductory Notes) may eliminate the danger of undue formalism.

G

General Archaic Method: Because many buildings are cubical, time tends to flow through the middle of a room, and more weakly from the tops and bottoms of corners of the room towards the middle. It has no additional proneness to pass along the axes extending from a doorway in the middle of a wall, but it does have less proneness to pass within small pockets in the corners of the room. However, the difference is sometimes made up by the amount of semantic time that passes. One will have a subjective impression of more time passing. One may feel alternately younger or more wise when sitting in the corner. Or one may develop a contrasting feeling between how one felt in the corner, and how one feels when leaving the corner. Whatever the case, the subtle trend is that time passes more quickly directly in the middle of the room. Since some of the time that passes has to do with other people's bodies passing through the middle of the room, an exception is found when a room is circular or has been empty for a long time. In these cases, time settles down, and there is a feeling of monotony. The risk in a monotonous situation is to spend one's own time un-thriftily. If one has a sense of boredom, it requires great wisdom not to lose track of time. This requires a great deal of mental integrity, which can be influenced by the cheapness or expensiveness of the building. Expensive buildings are more likely to feel monotonous for a long time, but they are also more likely to reward wisdom.

[1] _Method for traveling in old places._ Connect with memories you have had in an old place. Convince the powers that be that you are anachronistic (e.g. because you love the building). Travel years equal to the difference between your anachronism and your age.

[2] _New buildings._ New buildings are trickier. They exist in a pocket of new activity which defines itself to be new, and thus has abandoned some of the old associations of the building. Old memories are thus less effective in time traveling in a new building. The best bet is to have an important event which causes you to travel backwards to a time when the building was still new. This may involve preserving your memory of when the building was entirely new. Memory of the things that happened the first day you entered the new building can be used to locate yourself in the older time, the time when the building was first opened. However, finding a sufficient event may be difficult. Things such as political influence, inventing a perpetual motion machine, inventing a time travel device, or becoming famous can sometimes be sacrificed to reach the earlier time when the new building was completely new. It requires, as I said, a memory of how the building was on the first day you set inside.

[3] _Traveling to buildings that are neither young nor old._ This requires greater under-

standing of how time actually works. You may need to be highly inventive with your associations, and accept that wherever you travel to, not everything will be different from your current time. You may find that madness creates sanity in this scenario, and vice versa. You may need to become somebody who is 'lost in the cracks'. You may need to pretend to be a drug addict, take a disguise, or lie in your mind and say you are somebody else, somebody who lived in the earlier time. You can plan to travel to a similar place outside the building, but now you expect the place outside to be 'old and dusty' --- but instead of picking a future where it is old and dusty, you travel to the past, when there were dry leaves on the ground. You may even find in this scenario, it is easier to travel to a time when the old leaves looked young and fresh. Or, alternately, if its currently the Fall, you can pretend you're traveling to a future date when it is the spring, but really (secretly) intend to travel to a past date when it is spring. Influencing this one may depend on having some type of power of immortality, such as a personal principle to be old before you are young.

Glitch Travel: This is a good default option for thinking about time-travel, but it is not what usually happens. Glitch travel is when the entire world appears to travel backwards or forwards in time. Arguably forwards time travel could be explained by strokes, sleep or sleep loss, or drugs, or other types of cognitive dysfunction, so most qualified glitch travels refer to traveling into the past. This might be ex-

plained by such rare phenomena as authority-
--divine or otherwise, mental technology, sole
or exclusive existence, and meaningful priority. Less likely explanations involve divine
mistakes, space-time ripples, and collapsing
dimensions. Of course, in most cases, the
event is not real at all, but instead a more
common perception error such as an inventive memory or a dysphasia involving forget-
ting how much of the past has occurred. But
if noticeable events repeat, such as an alarm
clock, a routine activity which occupies much
of the time, or a political event, and you are
certain it occupies some of the same time,
then that is a sign that glitch travel has occurred. Someone may argue that it should oc-
cupy the exact same time frame, but that is
unreasonable, since the causal connection is
different. In fact, only certain events seem like
conclusive proof, and they are often available
only subjectively to the person who traveled.
Making time travel a historical event is a
challenge that has not been overcome, and
may have dire consequences if it was. On the
other hand, the simple concept of information
seems to allow for the possibility of a historical consequence for time travel. Simple but
deceptive, someone might say, and they would
erroneously conclude that time travel is deceptive. (These concepts contribute to another
section in this book called the Rhetoric of
Time Travel).

(The) Great Migration: The occult significance
of migration is summarized:

>   "The great migration occurs in itera-
> tions"

>   ---Eucaleh Terrapin

Migration is an old concept, but here it means
the levels beyond ordinary migration, the iterations of ordinary patterns that may have
higher significance. It is telling that each level
could be represented by a migration, because
this provides a dimensional framework. Depending on the quickness of the hierarchy, it
might lead us immediately to something pro-
found or otherworldly, or it might lead us into
mere translations of the mundane. Of course,
it is not the object to be unrealistic. Here the
second migration is defined to be time travel
of the temporal kind --- perhaps stretching
the imagination, but somewhat mundane if it
were accomplished without feeling. The first
level is ordinary travel, thus providing an extension through various metaphors of travel.
The third level could be immortal time,
whereas the fourth category is the timeless
world. These are really the only categories remaining to be discovered. Overall, the figure
of migration provides a convenient scale of
time-traveling phenomena::

H

Hard Semantics / *The Art of Hard Semantics* :
Several categories describe conditions under
which time travel is easier or more difficult,
or neutral to travel. This is what I call 'hard
semantics'.

1. An object is observed to be traveling more
slowly, and so, it does not oppose the back-
ward time-stream. The same object could be
harmful for traveling to the future.

2. An object is observed to be traveling to the
future, and so, it is beneficial to forward time
travel. The same object could be harmful in
traveling to the past.

3. An object is tranquilly in its place, like lem-
onade at a café. Such an object is not necessarily good for traveling to the past *OR* the
future, but may be used as a correspondence
with objects of the desired type. Some types of
neutral objects which connote power may be
desirable in spite of their neutrality, such as a
modern artist's paintbrush or objects made of
glass.

Health:

Modified Health Checklist for Time Traveling

A1: Drink fluids.
A2: Be safe.
A3: Have energy.
A4: Have protein.

B1: Strength / flexibility
B2: Eat nutritiously.
B3: Variation in diet.
B4: Diet / Eat fruits / nuts.

C1: Non-toxic.
C2: Healthy brain.
C3: Metabolic balance.
C4: Healthy bones.

D1: Moods
D2: Intellect
D3: Temporality
D4: Immortality.

Horcruxes / Incarnizing: This has a similar effect to 'saving the game'. It can be done at any point, in reference to any other point or set of points. The general theory is that its cumulative. It often begins with single, very casual points, and continues to more complex, more secure locations in time. The more self-assured points in which you have established a rational perspective, the better. Although horcruxes sometimes have an evil connotation, what I mean is something like a 'horcrux lite'. These are forms of preservation, like saving a game, that exist informationally. They are ways of supporting time travel without actually engaging in it. It may be compared to an engineer doing site-laying. The outside perspective granted from another position is designed to reinforce the structural integrity of the actual thing. They are essentially, self-officiated, official perspectives on your own travel through time. The idea of a horcrux or powerfully enchanted object reinforces this idea of a time in space that means something. The time in space becomes the enchanted object. The magic---if it is magic?--- becomes economical because the time in space is only encountered again from the atemporal or trans-temporal perspective. Otherwise it appears that it is discarded. Thus, the magic used to create the object becomes much cheaper to perform, because it only requires a certain amount of assembled information, and only narrow, specific, areas-of-effect. In general, this theme is more important--and has stronger effects---for those that plan to time-travel more than once. However, sometimes the first time is the most difficult to arrange, and therefore requires more site-laying. It may also help to have a metaphorical or side-real idea of the time landscape, to assist in site-laying. For this purpose, it may sometimes be necessary to have some authority, however secret, however ancient (e.g. from a past life), which serves to create the very information structures which are later used for site-laying or perspective-having.

J

Jettisoning: In the technical vocabulary of time travel, jettisoning is a bad thing. It is when the body is forced to move sideways. It is a by-product of high entropy or poor-quality material semantics. It can be seen in mentally retarded people: the willingness to shift the body left and right, for no apparent reason. That is not the only example. It appears in all kinds of people: in jolts, drunkenness, schizophrenia, predictable people, etc. It becomes important to avoid the symptoms if possible, because the symptoms show a willingness to discard aspects of one's reality without making gains in time. One strategy is to focus on the future (*perhaps that is the major thesis on jettisoning...*). Some kind of focus may be necessary. I have found focusing on the present to be very helpful in philosophical adaptation against jettisoning. If one can imagine a contingent space, a space in which one faces time from a sideways angle, this can be helpful in avoiding the harmful effects of jettisoning. However, if jettisoning occurs during time travel, the effects are less predictable. It may actually result in occupying a different world or time than predicted. The general rule is to stay straight even if you're going in a curve. That way you won't do anything you can't predict, like jettisoning everything. Hopefully that's impossible. By the way, jettisoning is a bad combination with un-attachment. Unpredictable, but true. The point of un-attachment is to find a route by which to travel. Jettisoning discards the route.

So, it isn't a good combination. A better combination would be luck or randomness.

Justified Time: As you time travel, whether it is in a machine, or through mental powers, you will increasingly find a need to justify what you do as you travel through time. For, without the proper justice, strange things happen. Trust me. There is a meaningful difference between justified time, and unjustified time.

1. What seems like the typical case of justified time may be more possible than other types. This is a case in which one travels through a (perhaps invisible) wormhole, and then later boomerangs back through the wormhole. This is a case in which considerable entropy was carried back through time along with you. Therefore, this could be considered legitimate travel, but it is not highly justified travel. Another case like this involved a time-traveling vehicle and magic powers. Although the magic didn't involve obvious entropy, and allowed a longer travel, it too had an effect of boomeranging back to the later time.

2. A second form that is also relatively common --- although still difficult, is time travel by officiation or 'assisted travel'. I have time traveled using the help of my mother and politicians. The travel that occurs consists of about ten days to several years, and does not require a boomerang. Because the power is justified by someone else, it is marked off as an arbitrarily justified power. Remember, the people involved may think you're worth abandoning somewhere in the cosmos, so the effect is actually somewhat deceptive. You

may actually end up in a later time that only happens to have the earlier date. Or, there may even be something seriously wrong with the place you are traveling to. But overall, it is not likely to be a very different place.

3. Another case is trickery, often involving the moment of midnight. This form of travel doesn't involve much entropy, so you may find yourself repeating the exact same event over again!

That concludes most of the direct experience I have had with time-travel! (One other time, around February 2010 - Dec 2009 I thought I had time traveled after entering a mental hospital, but I later discredited this due to my mental health, and because the only difference apparently was that my mother had cleaned my room).

L

Language of Time Travel: It requires a certain amount of certainty. For example, here is a short dialogue from once before I time traveled:

> >"It cannot be undone"

> <"I go"

The conversation took place telepathically, with an unknown power. There was barely a rushing sound, and I wound up in 1972 with the help of a woman. However, it turned out (as I suspected), that it could be undone immediately, and so I preceded to travel immediately backwards by force of entropy. But since I was such a cornflake in the language of 1972, I didn't suffer any damages.

Another conversation:

> >"CONTROL the FUN. Get what I mean"

> <"Time travel"

> >"You got it"

The occasion was high school. The senior class was taken unexpectedly to a fun-and-games area as a reward for graduation. To solve the riddle and win the prize posed by the mysterious figure, from an empathic, social point-of-view, I had to think around others' perception of everything. What are they doing when they implement something around me? Control. What motivates control, in all reasonable senses of the term? Fun (fun could even be semantic. The occasion gave me a hint at the answer---I had to believe in synchronicity). What was the logic of control? Perhaps government. What was the logic of fun? Perhaps freedom. Freedom I knew, meant volition, which meant breaking control. Breaking control surely didn't mean breaking government, so it meant breaking something over which the government had control. And I had to guess this was time, because otherwise the government wouldn't have REAL control. So, by the criterion of reality, we were supposed to 'control the fun' and what it meant was time travel---travel being a form of fun, and time being a form of control. I had won the prize. This was yet another occasion on which language played an important role in time-travel.

Another time, perhaps the most significant time historically speaking, was a time I went to one of those conventions that was mockingly arranged to receive time travelers. This event was in Boston around May 7th, 2005. I took the train from New Haven on a day where I would ordinarily have gone to class.

I waited and waited for someone to say something, so I could make a big appearance. "Who here is a time traveler?" said the spokesperson.

"I am" I said in a quiet voice. And I

raised my hand.

"Okay, in what way, or what devices did you use to travel here" he said.

"I used my mind" I said. "This is the second time I've lived or relived this part of my life". "The first time I didn't go to this convention".

"This is a village idiot" he replied. "Anyone else who has really time traveled from the future? The past doesn't count".

"But I really have traveled" I said.

"I don't believe you, and if you keep talking that way, someone will lock you up".

"Okay, well, I'll recant if no one locks me up. That's a good bargain".

Then a policeman approached: "Okay, come with me..."

"Just let me go this once! After all, you don't really know if I'm lying. I went out of my way to come here, so I must have some marbles in my head!"

"Okay, time travel now and create a double!" someone said.

"I can't right now. It's very difficult. And, there are no doubles..."

The policeman gestured again.

"Okay, I'm leaving, I'm leaving..."

The major features of this event were that I went out of my way to prove that I had time-traveled, and nonetheless I was ignored and scorned by the majority. And the denizens of this place tried to make desperate pleas to commonsense when I provided contradictory evidence.

Lemmetic Method:

Influences Upon Difficult Correspondence Between Time Dimensions

[2] Effect	[1] Thought
[3] Emotion	[4] Background

These are listed from easiest to hardest minimum,
and least to most common maximal effect.

Length of Time: Formerly critics argued that it is useless to objectify time: time exists independently, and moves at its own pace. The major detractor to this was the relativity and quantum view, in which perspective might in some instances come into play before the creation of the experience of time. So there is one view of time occurring statically-versus-dynamically, a kind of isometric time (the most traditional view, dating to Plato), and there is also another view which is more interactive, similar to cutting and editing movies. According to this second view there might be intermediate territories between static or immortal time and temporal or measured time. The traditional view might say we have come closer to immortal time. Modern critics might say that the current view is only a matter of semantics. But the question remains: can time be objectified? Or rather, can it be objectified in the same way as a work of art? One perspective is that this depends on two dimensions of time, which some say cannot normally exist. But time travelers would answer *yes*, there *is* such a thing as two-dimensional time, and thus, there *is* such a thing as time objectified as a work of art. This is a critical difference, which affects the perception of whether time is measured. The traditional view favors the idea that time exists only in a straight line. If there are dimensions in this view, they are branches from the main 'tree'. In the second view, the tree has ripples because of relativity, but the tree is essentially unchanged. However, some of these people think teleportation could take place from one part of the tree to another. This is the beginning of the two-dimensional view. Notice that the tree is already technically two-dimensional. But time is perceived as being one thing. The revolutionary view would be to revise time into more than one concept. However, this often does not occur, perhaps for convenience. The point is, in the time-traveler's view, the tree could be three-dimensional. Branches can loop, so that for-wards- and backwards- time might be the same thing. The radical answer at this point is a form of situationism: the idea that the cur-rent moment is a kind of 'hack' on multiple potentially relative or quantum concepts of time. This is the view that allows the time traveler to travel: multiple-relativity, and multiple-quantivity. Time is now measurable in terms of travel, and may take variables such as the time-travelers abilities, acceptable causality, and the amount of flux which takes place. Higher-dimensional time travel would probably secure multiple categories of vari-ables, and in these forms of travel the proper-ties of time would become more apparent, al-most like physical structures or boundaries of the universe.

(The) Location of Time: One should occasion-ally pay attention to whether there is a loca-tion to time, as this can offer clues as to one's level of reality, and opportunities for temporal power. In general, men can find time's loca-tion by avoiding masturbation. This will cause time to spiritually materialize in places like domed buildings, wind vanes, and hour glasses. Maintaining fear about the existence of time is known as spiritual dissipation, and it is that condition which goes unrewarded. Time exists no matter what (that is, for every-one who is not God), it is just a matter of whether it is *visible* or *manifest*. In the case of women, time exists in the mind or with God. It becomes the fundamental moral for women to equate goodness with intelligence, and in-telligence with the good. Otherwise they *be-*

come dissipated in time, as opposed to the male case, where dissipation appears as a doubtful source of motivation.

Logico-Matum: A form of time-travel occurrence. See under Matums.

M

Machine: See Time Machine.

Manipulating the Clock: Here is a technique used for manipulating the exact time of day. 1. Choose a particular time of day that seems reasonable. 2. Superimpose a different clock system upon the current clock system, citing your own prior authority. This doesn't in- volves any physical shift, unless... 3. Arbitrar- ily select your desired time in the current clock system as you would desire it to appear in the second clock system, appearing visually the same except with different numbers. 4. With some skill, you should then appear in the desired time, if you have not already looked at the clock.

Above: A graphic for manipulating the clock. See the alternate time system superimposed on the ordinary time system. Note also the emphasis on 'densities of time' indicated by the spiral and lines.

See also the section titled 'Naming the Time'.

Matums (Special Time Travel Occurences) - Meaning 'events' or 'mechanics':

Quasi-Matum: A quasi matum is the event which leads to an ALTI-matum I describe later. Essentially, one becomes more and more loyal to an alternate dimension of reality, such as by escaping the conditions of one reality, or through having no commitments to events that already happened, but which you don't feel like repeating.

ALTI-matum: I discovered this type of occurrence on 2/2/2015, around 3:25pm. My mother denied that one of our family tents had ever been burned, and there it was, almost as good as new, in front of me. I have a memory that the same tent got burnt when a candle burned low some years ago. It looks like the same tent, a characteristic brown and green tent with a characteristic triple-arching beam structure. And this tent is not new, it is just not burnt. So, what happened? Apparently, I traveled to an alternate dimension, where I didn't burn the tent! One way I might have done this is by time-traveling to a past where the tent wasn't burnt, and then not redoing the tent-burning event. This could be called an 'altimatum': a way in which knowledge of time narrows to a single event, and then deflates, as a different dimension is fully realized. Cool! The earlier part of the process is called a quasi-matum.

Logico-Matum. This process does not involve time travel so much as arranging your brain so that time travel could occur. It often takes place preceding an ALTRI-matum, and involves systematizing one's own value system so one can prefer one event over one or another.

ALTRI-matum. This is a variation of the ALTI-matum in which the new loyalty occurs instead because of a value system. Thus, one could to some extent be defining the rules by which the new world operates. Time travelers are considered high authorities about the laws of nature, as I have learned. ALTRI-matum can be defined more specifically to refer to time travel events in which one defines some of the rules of the new domain. Thus, they are similar to events in which a wormhole takes kernels of the former universe with it.

Parsi-Matum. A moment in which one identifies with a certain piece of information. The information becomes so important, that you attach your own entire significance with the existence of the object. This form of activity, also called objectifying, may assist one in time traveling. But it is not always certain what the expense is. For example, it is unlikely that anyone else considers the object meaningful. It is as if the object is animated to serve your purpose, and no one else's. The object becomes a specialist in 'you'. Simultaneously, one is likely to lose touch with one's own authentic past. If you are improving yourself, so much the better. But if you are disapproving, so much the worse! The object you use to objectify becomes analogous to the magical myth of the Bitter Cup: if you think one thing, everything gets better. But if you think another thing, everything gets worse. Some of the strongest activities with parsimating involve logical identification with objects, either through identifying appropriate objects to identify with, or through the use of an extended logic to modify the identity of the object so that it corresponds with other objects. Probably every mental time traveler has an intuition for processes like this, so what you read on the subject is likely to be less technical than the actual activity. But fortunately, life is open-ended, so creativity is a viable option.

Ulti-Matum: An ultimatum defines the base groundwork of the time travel continuum. When one has achieved the ultimatum (often in progressive stages), one comes to realize that one is in control of time and space. One acquires a sensation of having one's feet firmly planted on the ground, with a vision that life is an enormous white tunnel. I have not gone very far in this dimension, but it seems possible that someone who time travels enough would eventually become a god. Generally, the more times one has traveled, the more one is in touch with the ultimatum. It is worth noting that a lot of normal people, unless they are simply automatons, must have some access to this higher level of functionality already---although, why wouldn't they admit it? It is possible that some people have forgotten about their own powers, by forgetfulness (the sleep of Lethe perhaps?), or have come to attribute the time-travel powers to God, perhaps because of problems and inconsistencies which they have noticed, and only recorded in the language of panic.

Mereology of Complex Objects:

Perforations in metal (phone booth, for instance): Must be accommodated through subtle awareness.

Computers: Exist independently, but add complexity.

Mountains: Represent the eternal: they accommodate the traveler because they are changeless (relatively).

Consumer Advertisements: They affect information, when information is relevant to time travel (watch things get simpler or more complex depending on your interpretation).

Clever People: Either know about time travel or they do not. Ambiguous cases are like rocks in the river, and sometimes have to be interpreted. Ambiguous cases may travel with you by accident.

Magical Objects: Have their won set of rules they follow, either interacting or not interacting.

Wealthy People: May have influence. That's about all you need to know about them.

Designated Objects: The heavily prescribed

role of certain objects can make time travel more difficult, such as Zen rocks on a college campus. But these objects are rare, and can be overcome with intelligence. The major risk is if you are suddenly surrounded by a large number of these objects, such as perhaps at Times Square or Tienanmin Square.

Personal Objects: Other people's objects are ordinary, but one's own belongings are way-points and free-associations in the time-travel process.

Nathan Coppedge

Millenial Symbols:

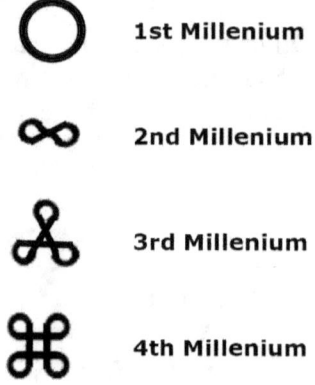

O   **1st Millenium**

∞   **2nd Millenium**

    3rd Millenium

⌘   **4th Millenium**

In a categorical system, further organization may be done by separating millenia into Ages;

Also, the symbols used may represent levels of a categorical system, in which every Age is a new application of the system; There are at least four alternate views as of the Third Millenium:

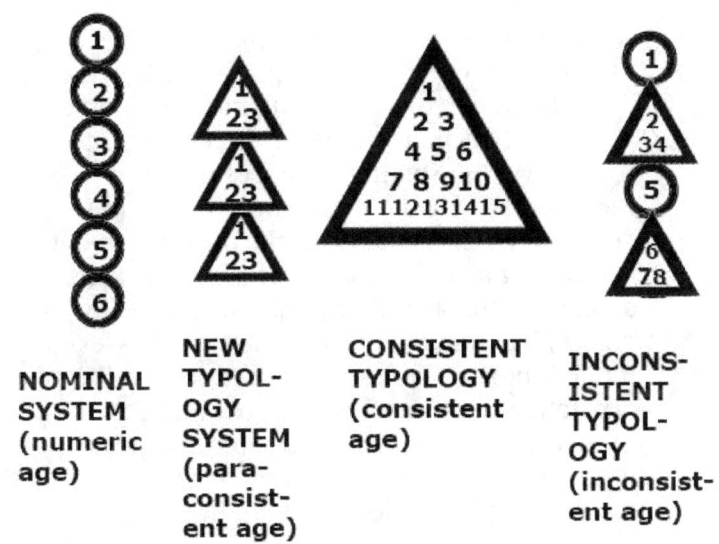

NOMINAL SYSTEM (numeric age)	NEW TYPOL- OGY SYSTEM (para- consist- ent age)	CONSISTENT TYPOLOGY (consistent age)	INCONS- ISTENT TYPOL- OGY (inconsist- ent age)

Respectively, the first refers to an economic concept of time, the second to a purposeful concept of time (entities and constructs), the third to an organized concept of time (government or municipalities of time), and the fourth to a purposefully determined concept of time (leadership, magic, or other asymmetries as might be present in an imperfect system)::

Modular Return: There is a kind of deep problem which emerges in the consideration of
how it is possible to remain synchronized
with loved ones. One solution is that everyone
has many manifestations or 'clones' --- any of
which is quasi-functional in the role of being
the same person --- with some subtle limitations. However, as the argument goes, people
are actually rather unique. This comes from
the theory that small differences in reality ---
the butterfly effect --- can cause huge differences down the road. Then the only argument
for finding a similar person is if very little has
changed. This would suggest that one really
doesn't have much influence over time---even
if one time-traveled. So there is a second theory, following from the view that differences
in time are subtle, and that therefore, any significant power over time would create en-
counters with people who were not adequate
clones of loved-ones. This theory is the theory
of Modular Return, which fits into the schema
of being a skill-based exception (skill-based
exceptions are analogous to applied theories
of morality). In this theory, some of the power
of time-travel is to return to the 'same place'
as before. This is quasi-independent of how
far one has traveled. One of the qua variables
is similarity to time. Thus, there is a kind of
empathic bond between the time-traveler, and
the context of travel. This can be explained as
the special socially- or technologically- con-
ferred status of being a time-traveler. It has
obvious limitations, which are essentially the
limitations of reality as we know it. The clear
point, however, is that time modules are the
perceptual framework involved in locating

sequences (which are locations). Thus, the problem in locating loved ones is really more broad than it was first conceived (at least in this system). Locating loved ones involves locating a sequence that involves one's own time-modular perception of the sequence. Thus, assuming that the other person has not time-traveled, locating that person is as simple as identifying one's modular perception of time in that sequence. So, time-traveling, in this method, is almost as simple as having the right module, which is not so difficult as guessing where someone is located on a complicated map. Thus, modular return has a potential usefulness in locating one's precise origin in time. However, if one cannot return to the exact time, it makes sense that one may not find the same self, or the same relatives, located there.

Morphization [Advanced Dimensional Techniques]: According to our biology, we have apparently two layers subject to change (the coelum and the dermis, which also means the coelum and the dermis of the head). If directivity other than inward and outward are treated as identical, and the motion is always planar (qua symmetric), then there are five typological categories of morphization. These are also conjunctive with whatever powers an individual may possess over the third and fourth dimensional planes (for example, 'wishing', 'free-will', and 'time-warping'). See the following diagram:

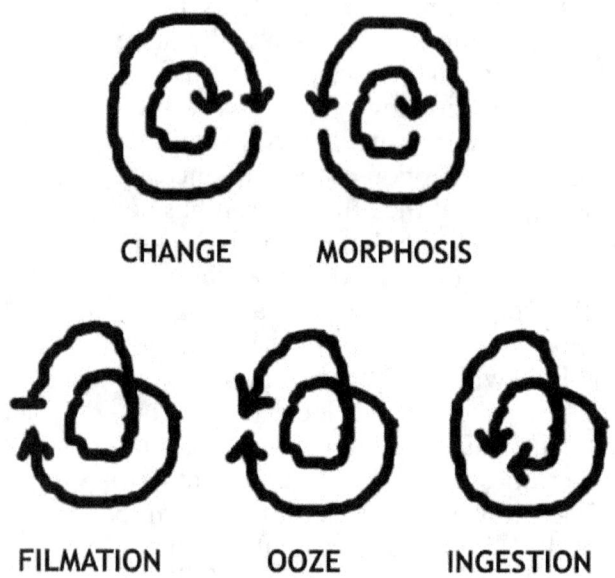

CHANGE   MORPHOSIS

FILMATION   OOZE   INGESTION

Change is the most basic type, and refers to 'falling through time'. The principle of mastery here is a simply abversion or contradiction, such as eating healthy food, 'waiting' in time, or in some other way earning time. Limits on the use of morphization make other methods look more appealing, unless there is a windfall of good fortune. This is the dimension with the most 'policing' preventing time travel.

Morphosis is more complex change. Travel occurs by contract with alternate dimensions, e.g. inserting islands of the temporal body ('islanding'). This is perhaps the most rudimentary---but also a rather difficult---form of dimensional travel. Because of the simplification of time dimensions, sometimes a very small amount of morphosis can be used as leverage for a great amount of travel, requiring subtlety.

Filmation is a process of simplification where one embodies and joins a time stream. This is a very risky activity, with lots of advantages, but taxes the body. It is a much more flexible rendering of the time travel format. It has a reputation for wisdom, but also historical accidents, according to the few observations on the subject available.

Oozing is a combination of morphosis and filmation: one embodies one type of time and uses 'islanding' to project and change into another type of time. This is like filmation, only more conservative, and more difficult, with fewer hazards. This method essentially requires greater knowledge of dimensions, but renders time to be slightly closer to the semantic, and thus much more flexible. Keep in mind that actually accomplishing it is more likely to be a matter of partial degrees.

Ingestion is not usually recommended, but can be used to ascend levels. Essentially, one digests the time stream, leaving one in a new place. The place is as new as one's ability to digest time is masterful. Having a new location in time can sometimes assist in acquiring power over previous dimensions, although at the expense of no longer being located in them.

N

Naming the Time: I would now like to introduce one of the subtlest methods of time travel, called 'naming time,' 'choosing time,' or 'selecting the hour'. This is a technique often using ambiguity or forgotten power to control the location of one's current time. Normatively, at least, these differences are trifling. They just involve a dimensional switch, so 'economically,' 'efficiently,' 'perfectly' the time is the same as before. The power is then to select an hour that is functionally the same, but actually different. The result is a kind of time travel, although there is not always proof that travel occurred. The primary proof that travel occurred is that one guessed the exact time that one is in. If cheating occurs against one's own time, there is no reason to know the time. So this is a relatively good defense against others' cheating with one's time. Here is a technique for this method:

[A] Establishing Knowledge
[B] Establishing Universalism
[C] Establishing Footing

These occur by mental processes. Footing is a technique such as reaction time, I.Q. , or better information. Knowledge involves some formal tool, such as one's best personal form of thinking, applied in an understandable way. It is basically a way of establishing that one could impress someone. For example, one could use 'painting by numbers,' knowledge of scalar enlargements, or simultaneous memory of a large number of facts. Universalism is about one's level of (personal) reality, such as how much of an impression life has made on oneself (even a memory of a yellow popsicle), or any claims to being a divinity, however dubious. With adequate establishment of the three factors, say, while blinking, it may be possible to choose the time of day very precisely, although it is more difficult to determine outside of one hour.

Notifications: Sometimes it is important to specify that a local entity is a 'spot reference' for the time travel event. This is especially true when the individual feels 'labeled' or 'tagged' by some other local entity, often acting on human premises. Sensitivity to tagging is a cultured skill, which often will only appear with a sensation of a mild undesired critical attitude. The actual source of discomfort may remain unconscious for some time. But a kind of ninja intuition to its significance can be developed without great consciousness of the tagger's importance. Spot references are important for several similar reasons: (1) As a common referent point between the present and the past or future, (2) To develop categories or words to use in the intellectual codification of time travel, e.g. in intellectual time travel, (3) To gain self-trust in the manipulation of time or space. Finally, notification is when, as if by magic, the inner process of time travel becomes codified through an outward coincidence. The simplest explanation of this is that an outward authority was notified of the intention to time travel---hence 'notification'---and vis. the tags this authority may also be critical or flexible. Consequently, notifications often occur before an attempt to travel into the past, and after an attempt to travel into the future. In this way, traveling into the future appears either psychic or pre-paratory (that is, occurring through an al-ready-existing future). Fortunately, time's natural motion permits this.

_P_

Panoptic Space: What would it be like if there were teeming people at every particular time, and in every particular place? No matter which way time proceeds, whether it is forwards, backwards, or in a loop, we can see that the people can walk freely from one time to another. There is continuity between each time frame. This is what is meant by panoptic time. There is no artificial barrier between one time and another. Each time exists in space. If there is something blocking the transit from one time to another, it is likely to be some form of physical object, or else the pace of time itself. Being able to view all of time from one place --- say, a loop of people, some traveling one way, some another, and some making it around the loop, many dying --- provides a sense of freedom and security. Although time appears in some negative lights, there are rules to follow, and the physics of it is not necessarily immaterial or vague. People don't just disappear, unless there's magic involved. Perhaps no one has a genuine *panopticon* of reality. But if they did, they would be able to gradually detect the best time-travelers, and perhaps even design reality so that time travelers could be saved with greater regularity. The Panopticon becomes the ecological symbol of time.

Parsi-Matum: A form of time-travel occurrence. See under Matums.

Placing Dummies: In order to clear up potential critical errors, it may be important to
place dummies. Dummies are concepts of
people, information, or events which conform
to the permissibility of time travel. Sometimes
a dummy is only opinion-news, a piece of information which someone assumed existed
because they found no contrary evidence.
Sometimes a dummy is a real thing placed
somewhere: a glove on a bed, or newspaper in
front of a house, telling someone that things
are going as planned. Even things which are
not necessitated by travel can be smootheners
for time: people will often behave very similarly, or at least in a better way, if they get
kind advice, or money, or a letter of thanks.
These things can also serve the purpose of in-
creasing positive entropy, that is, entropy
which permits travel to occur. In other cases,
a dummy might be a physical imitation of
oneself (or, more likely, another self from an
alternate dimension), which plays the necessary role that one can no longer play. Sometimes the dummy self might encounter problems that they are not prepared to face. Thus,
you must be prepared to accept a small
amount of responsibility about the fate of your
other selves. Sometimes, if the person is not
playing a meaningful role, and is actually
made up (perhaps by the government, or the
gods, who knows...) then that dummy self
may undergo death, and one's identity may
cease to exist in that reality. However, it may
be emphasized, that other people are often
more accepting of death than one may think.

It is not necessarily the case that other people will be miserable if you go missing. They may behave connected when you are alive, but when your death is certain, many people are more adaptive, and might be quite glad if they were imaginative enough to know that you are still alive. Perhaps the only people who are upset are those who are not seeking immortality. That's one way to put it. In general, dummies are less extreme than faking death. It is more like a medium of information which succeeds or fails in proving that the dimension is the same as it once was. The rule to follow is that if the dimension has changed, it may require some travel to get back. In general, the past doesn't change unless you rewrite it. This can be called the Rule of Fluxus.

(The) Pleasure of Time Travel: I have considered the subject of serendipity in time travel, and determined it is almost like its own type of travel. There is the usual warning not the confuse false events with real events. This simply says that real events are more rare than false ones. But if someone is completely unattached from reality, by accepting earlier or later conditions whole cloth, then there seems to be no psychological barrier to travel. What are the barriers on the time-traveler's happiness? They are several, and I will list them here: (1) The burden of being unattached to the past even when traveling to the future, (2) Accepting the emotional climate that one is in, even when it may be slightly unfamiliar, (3) The burden of history, that is, accepting the history involved in the places where one is traveling. The burden of history can be solved by accepting that one is only a small agent in the process. But, megalomania can lead to guilt in the process of time travel. Accepting the emotional climate simply requires psychological adaptivity. If you know that what you carry with you in your own process of history is very good, then you are less likely to have trouble with being attached to the original history. Sometimes un-attachment may actually cause you to grow close to your own story, because the contrast between the present and the past makes the past that much more clear. When there is a high degree of acceptance, serendipitous travel becomes possible, if there is significant technicalism, and the right conditions hold.

Power Majore Vs. Authority Minisculi: One of the major theoretical components to come up is the difference between force and authority. Force is often represented by probability, and the will of the majority. Authority, on the other hand, is something small and obscure. However, the two things can be mixed to-gether. One can have an obscure authority about power of the majority, and one can also have a power of the majority about *authority minisculi*. The result may be the ability to time travel. In this way, philosophical tools of knowledge, rare cases of being related directly to politics, and other similar unique advan-tages can be used towards the ability to time travel. It is only a matter of establishing a case and a rationale that is absolute enough to guarantee the case. When it gets to the point of ambiguity, the free will can be used to de-termine directly what occurs. It may not guarantee that those you know see the results, but someone may. They are likely to reject the reality of the matter, because they haven't traveled themselves, and they feel envious, but nonetheless, if time travel has occurred, time travel has occurred.

Pre-Meditated Time-Travel: There are a number of associated methods that contribute to the pre-meditated aspect of time-travel. These steps are not so different from magic. However, because we know dimensions of reality are real, and time-travel does not involve so much as a relationship with those dimensions, it seems substantially proven that some type of time-travel is easier than most types of advanced magic. Step one (meditation) is to focus on an object. The object may be as complex as the contemporary political scene, ideas of inter-planetary travel, concepts of allegiance, or as simple as the square pattern of the bathroom tiles. The focus should be on what these things mean _for time. Or, for the self-contingency._ Step two (metaphor) is to use the object of meditation as a symbol for time in general. Think of---what are my tools? In what way do I have a handle on time? Can I travel to the past or instead the future? Can I change time in small ways? What is my attraction in history (usually the current period, but sometimes there is a clue that some feature of another period can be found nearby---then, if you are attracted to it, you can try traveling there)? Step three (justification) is to have a reason to travel. This means abandoning all attachment to the current time period. Focus on an aspect of your object of meditation, an object you're attracted to, or an aspect of yourself that you do not want to change. Step four (alteration):

Do your best to imagine that thing or aspect in a different time. It may be important to do a lot of reasoning about motivation, to work up to this. How can you absolutely PROVE that you want to travel somewhere? There is a rhetorical aspect to this. In what ways are you like a person of a different time? This may be important: how would they travel? What can you do to explain yourself to a different time? (Shorter periods may be easier to justify if they are in similar locations. Longer periods may require explanations similar to insanity, or involving other people's idea of time travel). That's about it for this method. Some subtlety is required, but those four steps are the basics of one of the most advanced things that humans might do, outside of magic:

(A) Meditation,
(B) Metaphor,
(C) Justification,
(D) Alteration.

Primary Tractatus of Time Travel:

Part One: Easiest Method

1. Call the improbability of an object that still exists from an earlier time.

2. Justify that the earlier time was more probable than the later time.

3. With total commitment, travel to the more probable time.

4. If your current location did not exist at the earlier time, first travel to a location that did exist.

Part Two: More Difficult.

5. Otherwise, focus on the things surrounding a large object which might have existed at an earlier time. Use knowledge of the older things to reach for psychic knowledge of the older time.

6. Focus on how the old things are new at the older time. Ignore the things that only exist in the future. Travel to the older time by focusing on things you knew existed at that time.

7. If you cannot travel very far, focus on traveling short distances by using the same method. Typically, these will involve the same location and synchronized time of up to 40 minutes.

8. Use the method to learn other methods of time travel, such as specializations based on fashion, history, a focal object, etc. Then combine methods for greater effectiveness. Alternately, combine this method with another method you learn in this book, such as one of the Principal Methods, or one of the Four Methods, supplementing with the Divine Methods or Morphization.

Part Three: Historic Time-Travel

9. Obsess over some detail which you think was a specialty at a different time.

10. Make this small detail or 'nariety' a facet of your personality.

11. Clear away aspects of your personality that did or do not fit at the target time in history. It may help to have poor memory or a non-stereotypical insight into dynamic psychology.

12. Travel to the other time by finding an access to the justice of the nariety. This may involve the use of a magical helper such as a witch or someone who possesses a wishing ability. For example, when I traveled to 1972, I used a wish.

Dominant Time-Travel

13. May involve periods of rest to reduce entropy. Thus, patience is desirable. Mastery of some kind is always present in the traveler.

14. Time travel always obeys a rule of appropriateness of some kind. It may help to have strong reasons to time travel, or to know that you can return most conditions back to equilibrium. This may involve a kind of omnipotent reasoning.

15. In the long-term, or with 'multiple fluxes' going on, it may help to focus on events rather than physical objects. The events are more objective than the specific nature of one's environment, although clearly there is some environmental contingency in determining the exact *character* of events. Focusing on events will create intuitions about the rela-

tionship between past and future time-travel events. Events become time-travel events.

16. Ultimately, if you make time travel a habit, your friend is variability. You'll need to treat macro-variables like micro-ones. For example, you should treat economic stability like it's your morning eggs. If you're male, the instatement of a draft is like an attack dog on the loose. Regularly considering the conglomerated mass of these micro-macros will result in greater long-term survival rates.

Principal Method of Time Travel (84 Related Methods):

*Gain a position of rank and apply classicism and leverage;

> This includes the following submethods:
> 1. To be a god of politics;
> 2. To be a politician with orthodox methods;
> 3. To be a politician with advanced technology;
> 4. To be granted exception of rank;

*Retain a position of justified insignificance, develop a complex value system, and transcend by being tailored out of the system;

This includes the following sub-methods:
1. Ersatz mentality in a position of limited faith;
2. Linguistic ability without efficacy;
3. Subjection to pruning by acting powers;
4. Artificial interface assumption;

*Become a time-travel actor in a justified routine;

This includes the following sub-methods:
1. Be a principal actor under a time-travel politician
2. Have secret knowledge of a justified-insignificant time traveler
3. Be technically significant in an in significant context
4. Be classically significant in a significant time-travel context

*Develop a qualified time-travel significance;

This includes the following sub-methods:
1. Originate knowledge of popular or technical time travel;
2. Be highly unique in the claim of relating time and travel;
3. Have data which serves as a basis for an investigation of time
    travel;
4. Have reason to engage in time travel;

*Slide into time-travel from a related program:

> This includes the following sub-methods:
> 1. Learn teleportation;
> 2. Build a perpetual motion machine;
> 3. Found a government or civilization;
> 4. Be born into an immortal family;

*Attempt a bureaucratic time travel

> This includes the following sub-methods:
> 1. Master highly specific complexities of an office job;
> 2. Engage in those highly specific complexities;
> 3. Become forgotten to time;
> 4. Find a significance for re-emerging from the details;

*Childhood time travel

> This includes the following sub-methods:
> 1. Receive early rewards and education;
> 2. Adopt a pensive approach;
> 3. Find a significance for time;
> 4. Find the insignificance within the significance;

*Artistic time travel

> This includes the following sub-
> methods:
> 1. Obsess over significant location;
> 2. Identify people with location;
> 3. Find the people;
> 4. Identify art with location;

*Time travel by memory and location

> This includes the following sub-
> methods:
> 1. Identify a location with more than
> one time;
> 2. Travel to the current time;
> 3. Identify the location with travel;
> 4. Travel to the other time;

*Time travel by memory and thought, first
method

> This includes the following sub-
> methods:
> 1. Think about the location;
> 2. Think about the location changing;
> 3. Think about change;
> 4. Think about changing location;

*Time travel by memory and thought, second method

> This includes the following sub-methods:
> 1. Pick a random thought;
> 2. Identify the thought with the location;
> 3. Morph the thought;
> 4. Change location;

*Traveling by determinism

> This includes the following sub-methods:
> 1. Establish the knowledge of the past;
> 2. Establish a causal principle that would create time travel;
> 3. Cause the past;
> 4. Live the future;

*Traveling by volition, type 1

> This includes the following sub-methods:
> 1. Prove complexity;
> 2. Prove contingency;
> 3. Prove preference;
> 4. Prove volition;

*Traveling by volition, type 2

This includes the following sub-
methods:
1. Adopt a dimensional view of space;
2. Imagine that effects occur any
where within that space;
3. Reason that determined space is un
reasonable;
4. Establish a temporal machine;

*Divine time travel by significance

This includes the following sub-
methods:
1. To have proven significance;
2. To have dynamic significance;
3. To be granted dynamic exception;
4. To exercise significance;

*Divine time travel by mandate

This includes the following sub-
methods:
1. To have a proven exception;
2. To be granted exception;
3. To be granted exception to law;
4. To exercise the law;

*Time travel using magic, method 1

This includes the following sub-
methods:
1. To argue about mortality;
2. To argue that one is made of time;
3. To argue that one dies by traveling;
4. To argue about travel;

Dimensional Time-Travel Toolkit

*Time travel using magic, method 2

> This includes the following sub-methods:
> 1. To argue that magic is old;
> 2. To lose magic;
> 3. To gain time;
> 4. To gain magic;

*Time travel backwards using magical enchantment

> This includes the following sub-methods:
> 1. Imbue an object with age;
> 2. Get older;
> 3. Un-imbue the object;
> 4. Time-travel backwards;

*Long years using magical enchantment

> This includes the following sub-methods:
> 1. Imbue an object in youth;
> 2. Prevent the object from aging;
> 3. Later, imbue the object with wisdom;
> 4. Gain the youth of the object;

*Time travel using infinite contingency

> This includes the following sub-methods:
> 1. Establish an axis of contingency;
> 2. Assess: all experience is contingent, but not absolutely contingent;
> 3. Abbreviate existence as non-contingency;
> 4. Act so as to be contingent;

Principle of Everything: There is a principle borrowed from spell-casting and magic, specifically the magic of Guo the Immortal:

> "Everything around you is determined. When you are free, you can determine it. You have the power."

While this sounds like something from a cheesy Ninjutsu cartoon, it is actually a powerful principle, which has deeply influenced my ability to time travel. It can be combined with two other elements to equal time-travel abilities. These elements are: (1) Un-attachment, in the most anglicized way imaginable, simply not having an attachment to the events and objects in your life. This is better described as permanent un-attachment, almost a hatred for your life as it is currently lived. A willingness to live a different way. A willingness to sacrifice everything except the future of what remains. A willingness to sacrifice your own significance for everyone you know. From all you know, you might be dead. Or you might have died. That is un-attachment. (2) Historical importance, or some other motivator involving time, in order to be important enough to travel. Factors might involve political influence, relation to a news story (in my case), or careless disregard for a huge sum of money. Other examples might involve technicalities or technical knowledge or causal influence in the ultimate outcome of history, particularly including events such as the invention of a time machine, perpetual motion machine (in my case), invisibility machine, or teleporter.

Combining the three factors, including the principle of everything, one should be able to time travel if one is motivated, and has a good theory. Motivation and theory are simply a matter of intelligence and rational justification. Voila, improved ability to time travel.

Psychological Factors: The obvious factors to the majority are whether someone is indeed sane in the mind. I would say that contrary to some expectations, madness is not necessarily enough to prevent time travel. This can be figured very easily by imagining a machine for time travel. The machine takes little account of whether someone is sane or mad, unless the person is using mental time travel. And even so, the madness may involve the correct tools for traveling in the case of mental time travel. So, in neither case does madness prohibit the activity of time travel.

Other factors involve the psychology of witnesses. As I have mentioned under *The Language of Time Travel* heading, time travelers inspire a particular type of desperate common sense. It often involves an unwillingness to take any form of evidence as unqualified truth, e.g. because it is improbable. However, just because it is improbable does not mean that it is untrue. So, they are really performing a fallacy. Specifically, what happens is a certain form of 'time travel envy and rationalization' ('TiTE-Ration' for short). In this process, the envious person decides that the best evidence for the other person time traveling is if they, too, have time traveled. However, if the person is not envious, the symptoms do not in fact, take place. Instead, the person thinks that in general, nothing tends to happen, justified by the *emotion* that there is no reason to be jealous. So, in neither case does the psychology establish a reason to believe the time traveler, without 'going out on a limb'. Frequently, the people with TiTE-Ration will argue that going out on a limb is unreasonable, simply because it doesn't appeal to their psychological state. The overall maddening correolis is to simply disbelieve the time traveler, and if more evidence arrives, simply to declare that the world has descended into madness. Unfortunately, this is no more sophisticated than mere semantics, and does nothing to clarify the actual evidence when it actually arrives.

Another thing someone with TiTE-Ration might do, is declare that some sort of 'glitch' has occurred, perhaps even a mistake about information. While this is entirely possible when time travel has not occurred, it is irrelevant to the issue when it *HAS*. Therefore, once again, these people are performing fallacious reasoning.

It is unreasonable to expect a lot of evidence of mental time travel. Therefore, rejecting the few cases that provide any evidence is likely to eliminate all evidence that it ever occurred. Simply because the events are improbable does not mean that they don't actually happen.

It is tempting to cite the tight-jacketed mentality of science as a clue about the lack of such travel in their constituency. E.g. these people may be accepting the lack of time travel as an assumption, rather than embracing the unknown. Such an attitude ought to be considered the bane of science, but unfortunately, they find maturity to be a more foundational motivation than knowledge, when the knowledge may involve the unknown. Perhaps it is no more than a semantic difference, but it is one that may in exceptional cases, provide the basis for an empirical one. Simply saying that scientists are smarter in every case is as unbounded a reason as declaring that everyone is a scientist. If not everyone is a scientist, however, it is hard to see how science pertains to the universal. This kind of simplistic generalization that 'everyone is a scientist' is actually the enemy of genuine time travel endeavors, when it comes to realizing the highly exceptional nature of the occurrences.

Psychological Semantics: This is the opposite of so-called semantic psychology. Semantic versions of psychology have a negative reputation of glossing over morals and depicting life as insignificant. In the higher sense of that term, it qualifies psychology as a general discipline, but not the individual features of the person. Psychological semantics is different. It involves particular mental techniques: techniques involving flexibility and emotional maturation. Psychological semantics are qualified by activities like 'blossoming', 'realizing', and 'remaining open-minded'. These are techniques which are usually taken -for-granted in mental time travel, but they are nonetheless important. Thy are the types of things which may be of great assistance, but which have nothing directly to do with the method. Two approaches may be distinguished in relation to this type of semantics: 1. The hard technique, simply of applying pressure to existing systems and constructions, and 2. A soft technique, of becoming more mentally aware. Generally, this distinction can be drawn as a relation between mapping (soft structures), and faciliation (hard application). That is about all you need to know. Getting into the exact specifics of soft structures is not what this book concerns. However, a pointer can be provided that it concerns such themes as intuition, exploration of intellectual themes and themes of complexity, and idealization of the objects of experience, or some other comparable conceptual tool. So, there you have it: a concept of psychological semantics, or imaginative flexibility.

Q

Qi / Key Method:

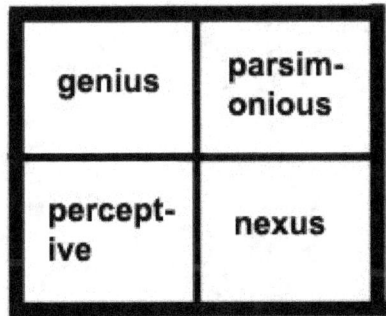

This method is designed to supplement more powerful or technical methods. It is a mind-focusing technique that helps to organize metaphysical knowledge. It is based on the theory of ambergris, which is rumored to be most effective in small amounts. Essentially, the most complete and perfect level is saved for the ultimate, and is left unchanged. Instead of acting on the real level defined as the ultimate, a map is drawn in metaphorical space which refers to local reality independent of the ultimate. Major actions are possible in the local reality by distinguishing it from any ultimate one. Intermediate levels are essentially ways of living an ultimate life, or ways of practicing spiritual materialization. These two final aspects are comparable with genius and perceptive aspects, which are distinguished by the exclusive context of this method, e.g. genius is innate, blind, and natural, whereas the perceptive is the spiritual body, the partially-exterior, and comes with difficulty. From an exterior perspective, the difference between these four levels may be blurred, creating power.

Quasi-Matum: A form of time-travel occurrence. See under Matums.

R

Rhetoric of Time Travel: *A kind of tractatus defending time-travel:*

1. Essentially, age does not prevent time-travel from occurring, as it is still possible to age gracefully during and after time travel.

2. Changing location in space is possible. If time and space are really a continuum, then it seems logical that time-travel could occur, because space-travel occurs.

3. Time travel does not depend on an exact time frame except to locate where it occurred. Thus, if someone is not concerned with the exact time, they only need general proof that it took place. Providing this proof, however, may be difficult, because the traveler is more likely to travel when he or she takes the exact same form as before and after.

4. Because dimensions of reality may differ in significant ways, the continuity of time in the place one travels to is not proof that time travel did not take place. It is simply proof that one is considered an odd ball. One is more likely to find evidence of time travel in the place that one has traveled *from*. Thus, it is possible that time travel involves some kind of rejection of an entire world.

5. Informationally, time travel may be permitted to have historical significance, although

people may have the convention of assuming it is a fantasy if they think gods are fantasies, and time travel is a god-power.

6. Although time travel exists as a kind of fantasy, this does not mean that it cannot exist as a reality. We know 'time' and 'travel' exist, and we know reality without time would be like time traveling all the time. So, really, the only thing standing in the way of time travel is the specific configuration of space and time.

7. The assumption that time travel is simple is often used as an explanation for how it is deceptive. In fact, this erroneously assumes that it must be simple. In fact, it probably isn't simple. Or if it is, then we might consider technology simple, too. In the abstract, if technology isn't simple, this doesn't make it impossible. Nor are simple technologies impossible.

Additional Points:

1. You may have to argue that you want to complete a real circle, even if it gets squashed.

2. You may have to argue that you want to meet the 'fake president'.

3. You may have to argue that you 'really want to connect the dots' vis. the circle.

4. You may have to argue that your life is not yet complete, and it is worthless compared to the ability to time travel.

Dimensional Time-Travel Toolkit

Rhetoric of Time Travel Vis. Immortality:

The moment one travels in is not dimensional.

If it were dimensional, then time would follow the dimension of travel.

Since it does not follow the dimension of travel, it must be 0-dimensional.

0-dimensional time is changeless.

Therefore, time travel involves infinite time.

Double-travel is double-infinite.

Double-time-travel is immortal.

Most time travel is not officially infinite.

Most time travel isn't official time-travel.

Most time travel is not officially immortal.

But time-travel is still time-travel, official or not.

Even un-official time-travel is unofficially immortal.

So I could be an unofficial god.

An unofficial god is not officially infinite.

But I can be unofficially infinite.

Immortality is semantic.

Time Travel is possible.

<u>S</u>

The Secret of November: I have found that April is especially good for local (short-range) time-travel of 1 - 3 hours. By deduction, then, November is the best month for long-term time-travel. Thus, it should be considered as such in any formal plans for long-term travel. For ease-of-use, all 'time-tables' for perpetual motion can be abbreviated as 'secrets of November'. Additional information: (1) 'Travel from November' can mean literally travel from November to an earlier or later date, or it can mean long-term travel from a certain date, (2) 'Travel from November-ish' can mean travel in which November or some other optimal date is in some way used to co-ordinate travel from a time that is not November, or a time that is not as certain. In other words, November-ish is used as a code word for careful uncertain travel, whereas November is used as a code-word for standard, well-coordinated travel. The distinction becomes important when a need arises to distinguish between time-travel typifications, and the specifications of actually doing so.

Secret Tricks: You can use time-travel for a number of tricky techniques, provided that you're careful. In general, a good rule is to do nothing you wouldn't be willing to risk doing now and then.

*Storing secret information: You can make statements 'off the record' and untraceably re-write files that you thought said the wrong things, using time travel. It is then as if the files always said what you recently wrote into them. This promotes a general mentality of high precaution.

*Historical influence 1: If you have the oppor-tunity, dropping your name earlier in the time-line can make you more famous, more ac-cepted, and closer to being 'your own parent'. As long as there's no reason your name would erase your name, it's pretty safe, compara-tively. You just have to remember that other people may not remember your youth in the same way you do.

*Historical influence 2: With a little clever-ness, you could make society more democratic (say, by raising the popularity of a motto which later became popular with the democ-ratic party), or some other group.

*Erasing your identity: If you live life onward from an earlier point that you traveled back to, this has the effect of erasing the history that happened after that time. If you want to go to extreme measures, you could take a dif-ferent name at an earlier point, and forget about your entire family. Although you would disappear from the later timeline, the earlier family would still believe that you lived out your life for a number of years with your for-mer name. This would give you time to sepa-rate yourself from your former identity. And

123

meanwhile, dimensions are likely to shift, making it seem like nothing unusual ever happened, because why on earth would there be two Henries, one who had changed his name, and one who didn't? The simple explanation is that you're a different person, as long as you don't travel back.

*Re-living moments: Perhaps the most famous trick is to undo a mistake that you regretted deeply. But if your mind is on time-travel, you might not be sure of the right thing to do. And heaven forbid, you do the same thing over again! The opportunity may come after a long while in the past, or it may come as an instantaneous 'wrap' --- whatever the case, the easiest thing may simply be to try to make a good impression. It usually comes off well!

*Additional thinking: Not only does time travel help the past, but it can help the future as well. Consider that if you travel to the past, that gives you more time to think about the future before it happens. This is an opportunity for virtue, I.Q., or strategy. Just don't be whipped around too much by the time pools! Find your center, and live in the moment! Unexpected things may begin to happen, eventually!

(The) Schmorgasbord: Before I developed any techniques (or what I thought of as my own techniques), I was offered---although I did not ultimately accept--- a Time Cube, which was considered to be a detestable sort of award similar to the Darwin Awards, for exceptionally anti-evolutionary activities. In the next few years after rejecting the award, I thought of the concept of the imaginary time-cube. Then I thought of an equation: 'real-imaginary time ^3' which I would use to prove the undisputable (partially) semantic nature of reality, and thus, gain access to time-travel. Another technique I used related to the time-cube was to deflect others' view of my local time configuration with thoughts and references to the imaginary quality of the time-cube. If others could be distracted long enough to think that I obeyed the rules of the time-cube, sometimes, perhaps, they could also be distracted for long enough to allow me to obey the rules that I thought the time-cube obeyed, namely, time-travel. This kind of double-semantics played a strong role, and was influenced by a conversation I had with my friend around the same time as the time-cube incident, in which I began thinking about the concept of 'over-clocking' or gaining computational advantages.

Semantics of Time Travel: Consider the numbers if time travel doesn't make a difference. This could be represented by the ratio 1/1. Now if we have two different time positions, then we have one difference. This can be expressed as 1/1 and 2/2. So, when we take these two positions, we can travel to ½, or perhaps to 2/1. Since 2/2 is mathematically 1, and semantically 2, when traveling to 2/2, it may be permissible to travel to ½ mathematically, and 2/1 semantically. We can never, however, travel to exactly 2/2 unless we are already located there. But this can be shown as the difference between having traveled and not having traveled. Therefore, there are some positions that we might be aware of:

Traveling from 1/1
[1/1 → ½ → 2/2 → 2/1 → 3/3 → 3/1 or 3/2]

Traveling from 2/2
[2/2 → 2/1 → ½ → 1/1 → 2/2 → 2/3 → 3/3]

Although confusing, there is some logic in relating the parts, specifically the compounding of previous travel destinations with the future number. For example, if one has traveled from points 1 and 2 independently to point 3, it becomes easier to travel from point 1 to point 4, and still easier to travel from point 2 to point 4. Traveling backwards through time, it is more complicated: If one can travel to 2/2 from 1/1, it becomes easier to travel to ½ from 1 or 2 in either end of the proportion. It also becomes easier to travel from 3/3 to 3/1 or 3/2, since 2/2 and 2/1 have been occupied.

Shaving: Elsewhere we have considered senses of the 'worm-body', the part of existence that appears to worm its way through time, and depends on time, or some higher-dimensional construct, for its existence. Now we will see that the worm body is modulated in a manner that depends subtly on the stature of the body. Shorter people are quicker and also more fragile. If we ignore that tall people sometimes fall from a height, height symbolizes further extendedness in time, whether it is a grandiose extension, such as a sense of largesse (as may happen with heavy people, which may be a sign of self-importance or else ignorance or despair) or else arrogant social prominence or some mixture of the two. However, the worm-body is not just the visible characteristics of the person in this sense of having stature, but instead the time-system as it imposes itself on the person. Conveniently in my view there is one primary characterization of the worm-body in relation to stature, and that is that the worm-body consists of a series of shelves, the largest one being the head:

The Worm-body

With that much said, there is a technique called 'shaving' which is when the lower shelves are made to line up with the upper shelf, producing a temporal body that is highly mentally responsive. While this may adduce a high degree of sensitivity (e.g. to mental stimuli), it also is a highly effective model for one form of time travel, which is the travel which occurs when one chooses what precise time it is. Conventionally this is understood simply as 'knowing the time'. But in the wider range of experiences involved in time-travel, it necessarily means being able to *choose* what time it is. So, there you have it. Aligning the shelves of the temporal body into a forwards, flat surface results in knowing the time, or in other words *choosing* the time. And this requires knowledge of the worm-body.

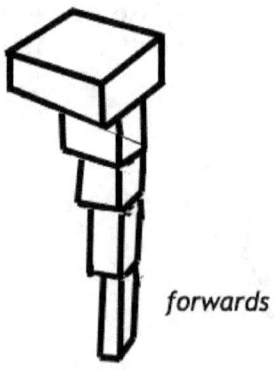

forwards

Shaving Using the Worm-body, a technique that allows one to "*choose*" time.

(The) Simplex and the Complex: ('simplex' means a simple entity, 'complex' means a complex entity). As it turns out, reality as we know it is only about as complex as the demands set upon it. This involves a radical technicalism, but it also unfortunately involves a high degree of functionalistic thinking. The resulting minimalist concept of reality can be called the 'Simplex' --- it is not so different from a thick slab of recycled paper. The paper can have many properties, colors, maybe even schema (if it so happens that some of the content is recycled newspaper). This may be an exaggerated metaphor, but it

128

is not beyond the imagination of a genius. A genius can imagine that life is simple. The point is, you can choose what seems complex about reality, once you apply a few variables to the competition. A physics lab, for example, is being very 'discerning' (physically discerning). It is using a lot of energy! What is the competition? Compared to the physics lab, most things look simple. A few transistors. They work in a certain way. They are not always relevant to time travel. Ask the transistor, want to contribute advice? --- or etc. etc. The point is, the person who wishes to time travel is a kind of exceptional system, a 'Complex'. Compared to the Simplex, the Complex is more dimensional, because it is concerned with more dimensions. It is possible to be more technical. When it is stated imperatively enough, then travel is granted. The Simplex --- the fabric of space time or your local gardening club, etc. is simple enough that the Complex of time travel can overcome its barriers --- rarely perhaps, but it is possible.

Six Bases for Time Travel:
[Important foundational concepts]

Improbability Drive -- Douglas Adams' concept has lessons to teach. Specifically, there is a relation between space-time, technology or technique, and risk or genius, which is the basic formula for time travel. Although time travel does not aim to be 'true random', randomness is clearly a variable involved before

129

travel takes place. The concept that someone might benefit where one is traveling suggests a karmic rule for time travel as well, that is directly related to improbability. This is the 'mysterious' dimension of time-travel.

The Way is Older than God -- This saying from Taoism is the democratic secret of time travel. With sufficient genius, every person has access to laws deeper than the surface of reality in which we find ourselves, older even than the authorities which think they under-stand how reality works. With this discovery, we have trump---an ultimate trump--- against the laws of physics. It is encompassed in the rule of inexorable, gradual change. Since time-travel is temporary, it places no complete stringency on the laws of physics. Therefore, it is physically possible where there is no net change of energy. This is the 'spiritual' dimension of time-travel.

Epochranysm -- Extreme concepts of energy create relativity flux, permitting exceptional movements. One has to do little other than create a time crystal in a highly realized di-mension that has a lot more energy, and one can then use the time-flux to travel within lower dimensions which influence the crys-tal's composition. I feel that I have done this once. It may have even been an important as-pect of my mental time travel. It essentially involves imagining surviving an atomic blast, and then becoming excited about what it means for relativity. However, this entire process can be abbreviated as the 'energy' di-mension of time-travel.

Catalyzing a Leaf Turning in the Wind -- Elements of time have long meant elements of change, which were represented in visions of chemical change. Chemistry became the symbol for change, and thus, time. However, chemistry was essentially a physical state, leading to the conclusion that time had been reduced to a material object. Although this corollary was often missed, it is represented in the concept of catalyzing a leaf blowing in the wind. The symbol represents some formal properties of time travel, such as resistance, permanence, and materials modified by time. This motif can be used as the 'tricky' dimension of time-travel.

Flux Bux -- The concept of a time-continuum expresses an idea of a universal currency which exists regardless of whether matter is present. Thus, controlling time is potentially more powerful than controlling matter or chemistry. Time can influence the composition of matter and chemistry. Time is a universal currency which can have properties even where no other properties are apparent. This is the 'economic' dimension of time-travel.

Formal System -- The properties of time-travel are not just the dimensions of time, but the dimensions of objects which exist IN time. The exact shapes of physical objects, the amount of energy present, etc. has subtle and sometimes not-so-subtle influences on the potential to time-travel. It may limit the types of time-travel that are possible, or the characteristics of the time-travel experience, or the distance that can be traveled. Larger objects, for example, pose a larger distance of travel, and thus a larger risk to traveling. Overall, the effect of physical non-time objects is the metaphysical dimension of time-travel.

Slide Travel:

"It's a general property of nature, like so many other stupidly brilliant but failed concepts" ---Perinnes

Slide travel is a problematic form of time travel that is known to happen sometimes through divine intervention. Essentially, clacking plates in an alternate dimension of physics (call it the 'slide' dimension, because that's essentially what it consists of), click together, extending in a long line, and projecting the traveler well beyond their expected destination. The effect is not any ordinary sort of aging, but instead a logistics error. The problem is compounded if the god intervening is luckier (as often happens) than the person traveling, meaning that the luck alone may have been the principle causing the slide. I typically experience slide travel at times when I feel moderately good about my life, and the result of the travel is to feel somewhat worse, and lose contact with people. In other people's cases, the travel may occur for other greedy reasons, such as divine intervention to lose business opportunities, or to increase the likelihood of a divorce. Slide travel is roughly the largest folly that can occur in time travel short of personal / physical damages.

Slim Picketings (Parallel Worlds Kinesthetics):
One of the most difficult aspects of time travel
may well be the ability to develop a map, a
map to move in the first place at all, and secondly a map to justify that time understands
that something has occurred. For the justification by time for nothing at all is a threatening
force of nihilism for the time traveler. The impression of travel creates a semblance---a no-
tion---of a secondary turning point, which is
like another being or system, called time.
Time casts its ropes around the world, and
people can sometimes move between them.
But when ordinary travel occurs, the ropes do
not appear to move. Lifting the ropes only
seems to occur by complete morphization---
resembling death, or abandoning the world,
or taking part in a divine mythology. But
within the context of experiences, time travel
may occur by setting the ropes oneself---for
the ropes are very subtle. They are described
as the hazy limit of objects, and the causal
axes of reality. It is not effected by becoming
unconscious, although there may be an impression of a cloud, or light coming from
clouds on the horizon. Signals like the following have significance:

Large spots of light (distant):
    Need for subtlety.

Dark pool, cave, or shadow   (forwards):
    Need for time.

Vertical objects (variegated in height):
    Need for complexity.

Central motif (spoken, visual, or musical):

Need for action.

Feeling of large, familiar objects:
    Resolve against travel.

By making these decisions at these times,
ropes are set on the horizon, enstabling the
ability to move to another dimension. When
that happens---this can be tested gradually by
maintaining a calm confident, artistic approach, or whatever mood suits your approach to life---then you must apply a time
travel technique, such as one of the Four
Methods, or one of the Eighty-Four Methods,
to ensure that travel occurs. There may often
be a sensation of placing ropes upon the new
world, which then provide some degree of
priority or character for that new context in
regards to time travel. For example, it can
sometimes be observed whether the old conditions degenerate, or are instead accepted or
enhanced by the new environment. For example, the new context may be more political,
magical, or conserving of powers.

States of Awareness: These can relate to the
worm body and other aspects of physical time
-travel.

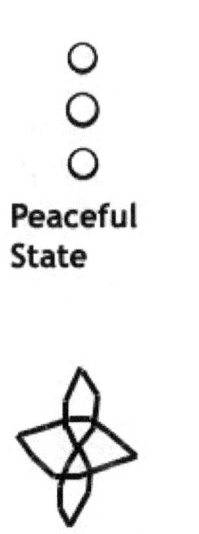

Peaceful State

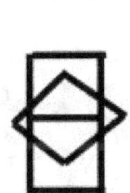

Aware State

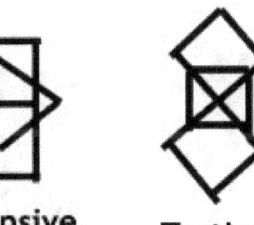

Pleasure State

Atemporal State

Defensive State

Tactical State

Doing: Past

Knowing: Present

Seeing: Future

Symbols : (See also under Millenial Symbols, and the symbol I crafted myself after the introduction)

SYMBOLS THAT COULD SYMBOLIZE TIME TRAVEL

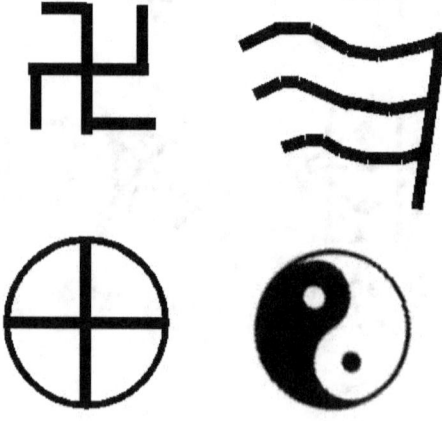

T

Taste Tactics

One critical technique may be to tempt your-self to time travel. Indulging an abstaining in the sensations of each given moment from past and future positions can motivate you to travel to different particular times. Here is a group of techniques:

1. To attract travel to the current moment from the future, indulge in sweets.

2. To make travel more complex (such as to have references to think about for traveling to multiple time-periods), indulge in sour and savory things.

3. To navigate to a given moment, ignore the present except in the most selfish and thoughtful way, and along the way to the past or the future, convince yourself that it is re-lated to further past or further future occa-sions on which you indulged in particular tastes. Note also that travel to the future may require knowledge of the future, and travel to the past may involve a vivid memory of how you felt at that time, in relation to other peo-ple, places, and things.

4. To avoid attracting travel to the current time, avoid sensory pleasures. This can also have the benefit of making travel across that region easier.

5. To have more time, think about the flavors that you have indulged in the most. This way the unknown future will seem less significant in its pleasures until it arrives.

6. To carefully prevent traveling too far into the past, deny yourself thoughts of the pleasure you originally associated with the time you were traveling to.

That's about all for taste tactics!

Temporal Mitosis: This is a bizarre, miscellaneous concept to consider. It has both positive and negative qualities. It is the ability to be located in multiple places at the same time. Often, the stretching into one or another dimension will involve a new feeling, which now pervades your entire reality. This is the feeling of mitosis. The feeling of having one, comprehensive, although perhaps flawed existence. For example, I have found that joining with the distant future creates a boredom, depression, and a need for frenetic activity. Joining with the past creates happiness, but also a deep-set laziness. So neither form of reality expansion is precisely good for you. My explanation is that both involve time. It is really through a-temporal mitosis that positive attributes might be had. But, if there is something wrong with development in the temporal world, there might also be things wrong with the a-temporal. For one thing, it is likely to be more challenging. And change does not occur very naturally without time. There is then a likelihood of some kind of sacrifice, fate, or mission to achieve development outside of time. If there is something encouraging about all of these observations, it might be that the temporal mitosis begins with laziness, so that means that the a-temporal mitosis might not be impossible.

The Thresholds of Time Travel

1. Freedom: Freedom is necessary in the early stages of time travel, because without freedom you will be unable to cross the mental and technological boundaries to reach a new dimension.

2. Insanity: The second threshold is insanity in the sense that you need to embrace radical new experiences. Also, the radical, blind (although finite) freedom of time travel can lead to a less rational view of the world.

3. Reason: The third threshold is difficult to cross, because it involves extreme reasonableness about what is meant by the freedom and madness of time travel. Crossing this boundary creates greater control and progress with the mental and technological boundaries.

4. Spirit: Finally (at least in my current view), beyond reason there is a spiritual barrier in order to embrace the full knowledge of time travel in all its dimensions. This knowledge is not just a technical knowledge of the world, but also a form of self-knowledge which has a recursive relationship with events in the world, including time travel.

Time Delay: One subtle factor to notice that may influence experiences of time travel destinations, is that there is a tendency to travel to neighboring dimensions which experience a slower effect of time. Tentatively, this means that life may become less challenging in these neighboring areas (even inexplicably), or even, perhaps, stupider. For whatever reason, many of these alternate dimensions were given a bigger break when it comes to time resources. Numerous theories impinge on this, such as Murphy's Law stating that the worst things happen to the best people, and perhaps the level of critical awareness effects one's baseline level of performance, or not. Whatever the case, there is a diagram that can be formulated which shows that depending on how far one travels, the place one travels to is likely to be equivalent to one's original dimension, plus minutes, years, or centuries of 'slack time' --- meaning the dimension one travels to has had a much easier time getting to where it is, differentiated by the amount of time one is traveling. There may be other cases where intelligence or resourcefulness creates a 'niche' in the new time frame, but for most purposes this is an extra-magical ability, rather than a standard feature of travel.

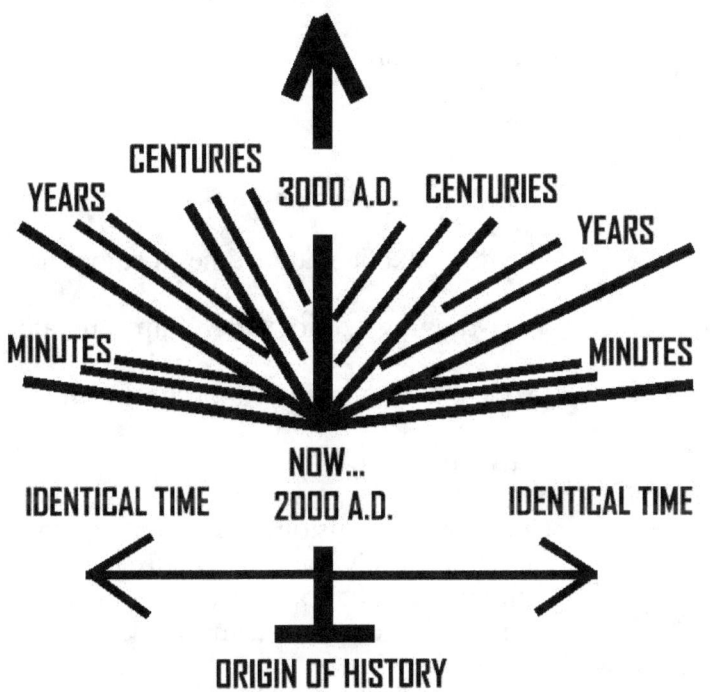

TIME TRAVEL MAY INVOLVE ACCESS TO DELAYED DIMENSIONS

CENTURIES
YEARS 3000 A.D. CENTURIES
YEARS
MINUTES MINUTES
NOW...
IDENTICAL TIME 2000 A.D. IDENTICAL TIME
ORIGIN OF HISTORY

Traveling longer distances tends to permit access to lazier dimensions. But it may also change how reality works! So there is a compromise between being well-treated and not knowing the laws!

Time Mahine:
How to Build a Time Machine:

A First Suggestion

Components:

*Local matter deep-scanner.

*Quantum exotic matter time- locator and

*wide-ranging exotic matter map generator.

*Exotic matter generator / programmer.

*Seats / booths.

*Universe / dimension synchronizing tool /
computer, mostly for making sure people and
equipment can be translated as exotic matter
from inertial reference frames.

*Event initializer / time editor / opportunity
locator / computer for locating or else creating commonalities in timelines for the purposes of synchronization.
I'm not sure you could take equipment that
way, but you could try.

By the way, I have no direct experience with
time-travel equipment except my own mind
and perhaps magical people or disguised ve-
hicles.

All of the above is merely an interpretation.

Dimensional Time-Travel Toolkit

Time Recapture Technique: A technique can be developed to waste time and then slow-burn it on long-term health benefits. The first time I tried this, peach-fuzz regrew on my face (I had already grown a full beard more than once).

Time Traps:

Schizoid Trap - This is one of the most common forms of traps encountered in time travel. If one cannot find a continuity or connection between one time and another, then time travel is not likely to occur in the first place. So, this trap is marked by an inactivity in time travel.

Boundary Trap - Another case is a boundary problem: when two worlds are significantly unrelated. This is a trap in which it takes a miracle to bridge the gap between worlds. A common property of this type is 'small similarities, marked by a definable character'. Achieving the capacity for similarity between worlds, while also being free from highly technical definitions of neighboring worlds is what could allow time travel.

Relativity Trap - If one time-travels for long enough, there is a possibility that one will become radically isolated from others, an effect called prolonged non-identification. This effect is attributed to the interaction between one's own time flux variable and the capacity to receive such variable by the ordinary time stream. While a relativity trap might be a key to achieving magical miracles (such as, notably, melting snow), it also has a downside in that it may cause PNI.

Information Trap - Time travel sometimes
seems to reduce to pieces of information
which permit small or large changes to occur.
However, if one doesn't recognize that time
travel can occur sometimes by information
alone, then one is consequently stuck in an
information trap whereby one cannot relativ-
ize the difference between ordinary time and
traveler's time. This is a small detail, but it
sometimes makes a big difference.

Time Travel Capsule: Ideally, a time-travel
module would have a number of technologies
or capabilities:

(1) A recorder and re-tracer of past history,
however crude (a video tape might sometimes
provide substantial evidence, unless the video
recorder does not travel with you). There
should be a capacity to make notes and mark
doubling-points. A constantly running video
or set of memorized notes may be adequate
with mental time travel.

(2) A predictor of future history, in case one is
traveling to the future. One thing to guess
about is the effect of one's travels in the past.
Probably the effect on the future is something
like the government's attitude towards perpetual motion, such as welcoming, or greater ob-
session with mental illness.

(3) A home base. Perhaps there should be a
particular place and time that serves as a base
to return to. Returning to base can clear up
things like 'do I have multiple bodies' (my
guess is 'no' due to the exceptional rules) and
'what has really changed?'

(4) The ability to regain youth in order to prepare for long-distance travel. This would be one of the remarkable health features of a genuine time machine. However, in my experience, the time is always lost. The only way to regain youth might be to become healthy, rather than to become young, unless you can morph your body.

(5) The ability to extract time from objects, such as coins and jewelry. Time might be used as a raw resource which permits freedom of travel.

(6) Any technological capacities to permit forwards and backwards motion. I consider mental time-travel more viable. However, technology has more promise with videotapes than mental events, due to the bulk of the machine, and the fact that objects can travel with the person rather than reverting to their earlier states.

(7) Perhaps the ability to convert the time capsule into an abeyant time crystal would be helpful in averting disasters, such as by avoiding dangerous events.

Time Travel and Depression: People encouraged to not feel depressed often gradually feel less depressed; Only, this tends not to be true for people who are unrelaxed; If being unrelaxed is a symptom of poor temporal powers, then there is some significance to the statement that people who do not time travel are depressed.

Time Travel Energy Concepts: Theories such as relativity, mathematical coherence, and reverse entropy may have bearing on time-travel. Specifically, these theories relate through concepts of energy. In particular, time travel takes a variable (energy), and applies the variable to a special condition, which is whatever theory applies to energy. So, for example, in the case of relativity, what may take place is being fast enough to be beyond the arrow of time by some definition. In a multiple-dimensional universe, this may even take place at sub-light speeds. It is a matter of traveling the equivalent of light speed between dimensions. This can be defined in terms of the physical distance between the dimensions, allowing mental time-travelers to 'cheat' by free-associating about their own lives, in spite of the supposed distance that time has traveled throughout the duration of the life. So, for example, if one lives for 30 years, and wants to time travel to the beginning of one's life, one would need to free-associate the equivalent of 30 light-years of time's arrow. But to travel half that distance would only require 15 light-years of time's arrow, and would only require a dimensional differential of ½ of the life that had been lived. Thus, in theory, if one lives for 30 years, one could free-associate at half light-speed and reach 15 years time-travel into the past. To time-travel only two years after fifty years of life thus theoretically requires free-associating at only 1/25th the speed of light. See Appendixes for further data about free-association. Loss of energy from aging could explain why people sometimes consider time-travel to be an activity for young people.

Time Travel as a Survival Technique: I remember one occasion on which I used time-
travel as a survival technique. I had fallen
from a five-story apartment building, and
while falling developed the concept of re-
spawning on the ground. This was accomplished by time-traveling upwards through
my body, until I created a kind of time-crystal
with my biological matter. This happened
automatically once I was able to time-travel
backwards that distance (about 25 feet), because the forwards motion of the fall created
an automatic loop with all the time-travel positions (or perhaps more accurately, points).
What was important was to have an objective
view, and to keep in mind that I wanted to re-
spawn on the ground. I even kept the word 're
-spawn' in my mind while I fell. And I didn't
wait until I hit the ground to begin time-
traveling backwards. The idea occurred to me
for the obvious reason that I wanted to go
back up to the ledge where I had fallen, since
that seemed like the only safe place. But it
turned out, I could survive by creating a time-
crystal. I wouldn't recommend this as a consistent technique, as it is highly contingent on
the level of quality-of-life that is ongoing at
the moment. It may also require a nearby divine presence, and concordant approval (the
term 'concordant approval' I associate with
this specific event for reasons I don't completely comprehend). The spell Frizzly Few 6 -
-- perhaps guessing the name of the world I
was in --- may have figured into my recovery
from the fall.

Time Trouble - Reflexivity: Problems with
time travel are generally of several kinds: (1)

Accretion, a technical word for complexification and computation requirements (ripples,
which are not necessarily visible in the best
cases), (2) Dimensional problems including
loss and gain of time, space, and place and
monetary or political resources---the so-
called recourse problem---and also over- and
under-extension problems, all mitigated by
the urban institution or bureaucratic methods, and (3) Flavor variance, such as not finding the same items for sale, and adjusting to
cultural differences (or eon flux); In general,
these will be mitigated by the natural synchronization between the world and the traveler ('say goodbye to reality' / however
gradually); Inhabitants are likely to consider
some adaptations to the traveler to be natural,
even from a passive position; Thus, there is a
fourth problem, (4) Reactive environments;
This is likely to be a problem mostly if the
travel is artificial and interpreted as expensive
(e.g. government deletion, habitat elapsion,
two things that can be controlled by thought,
assuming there is a choice of destination);
However, many of these aspects are prospective boons: e.g. government provides for time-
travelers, citizens / denizens attracted to behavior, or new habitats built with the intention of being located; Thus, the fourth stage
actually tends to be more comforting tha the
general complexification / computation requirements, at least in the case of mental time
travel; One major concern to address is the
balance of whether time travel only occurs
during periods of youthfulness, and if age can
be compensated; Clearly, the time traveler
must aspire to immortality unless life is
sheerly objective::

Travel Patterns ~

BACKWARDS CONE PATTERN IS NOT IDEAL, BUT IS ANTI-ENTROPY

THE PHOENIX PATTERN IS POTENTIALLY LONGER-LIVED, BUT DIFFICULT; TENDS TO INVOLVE UNIVERSAL HISTORY

THE FORWARDS ARC MAY BE MORE ADVISABLE WITH POLITICS OR ECONOMICS THAN QUOTIDIAN SIGHTSEEING; IT INVOLVES INVESTIGATIONS AND RECOUPING AND TENDS TO BE DEGENER-ATIVE

Trinity of Methods ---The following method of 'scheduling' may be preferred as the primary trinity method. It may assist in both mental and technological time travel. A later method that follows is more general.

Method 1 / Macro Method:
Date: Changes
Hour: 12am ~ 12pm or other reversals
Year: Changes drastically

This method uses contrast or physical shifts in
time structures to create an 'epochal difference'. Little caution is taken to the effects of
travel, and very often everyone in the world
travels at the same time. It can also be called a
'dimensional shift' or 'dimensional gambit'
similar to an instantaneous eclipse of time.

Method 2 / Medi Method:
Date / Hour / Year: pick one

This method uses a rule of similarity between
the points of travel to create a 'synergasm' or
invisible wormhole between the two involved
worlds. Like the first method, there may be no
major similarity between the two correspond-
ing worlds other than a temporary temporal
similarity. It may even be argued that all temporal similarities are temporary.

Method 3: Micro Method:
Date: Later to earlier
Hour: Usually 12:01am - 11:59pm approx.
Year: Generally the same

In this method, the hour is chosen paradoxi-
cally so that generally the date changes but
not the time. Since the division between 12:00
and 12:01, or 11:59 and 12:00 is infinitely
divisible, so too, there is an ambiguity in what
date of the year it is. Consequently, sometimes
a time traveler can choose to be located at
precisely 12:00, but moving slightly backwards through time, thus achieving time
travel.

These three methods may be used as a simple,
although important alternative to the Primary

Methods.

For the most general and abstract methods, a method mentioned under Corporate Dependence may be elected. These represent potential relations between matter, time, and energy necessary for time-travel:

> 1. Between matter and matter vis. Time,
> 2. Between matter and Time in terms of space-time-energy, and
> 3. Between Time and Time, in terms of matter.

U

Ultimatums - A form of time-travel occurrence. See under Matums.

Unitation - Is the general continuum of time, according to atomic perception. With a honed ability to perceive the continuum, discrete time units become visible to the mind (usually as a 'force' which is invisible yet palpable, which exists simultaneously yet separately). This ability becomes highly important for many types of mental time travel, and also technological travel, so far as it can be intellectualized. It is possible, according to this atomic view, to treat time units / clusters / globules as islands which have a mysterious function. The first insight on the subject is that some globules are performing a negative function, and some a positive function. Positive functions grant additional time, although they do not always prevent forward motion. Negative functions often drain the remaining time*, however, in relation to the positive elements they provide flexibility which increases the speed of travel. According to my assessment, these observations are not superficial, although they may depend somewhat upon factors like one's level of semantic knowledge, differences in natural law, and physical proficiencies or dependencies.

*Although according to the theory of amber-gris in the following volume, sometimes subtle negativity of these time units is more useful / flexible than the exaggerated positive, because of unpredictability. This is the true intellectual or supernatural version of ambergris, rather than the narcotic one.

Dimensional Time-Travel Toolkit

APPENDIXES

Nathan Coppedge

APPENDIX I. QUESTS
(Whatever you do, don't kill your grandfather!)

BEGINNER QUESTS
Study Philosophy and Metaphysics: This will give you an edge in your knowledge of time, destroying others' ability to second-guess your time-travel attempts.
Know the Basics of Relativity and Quantum Mechanics: These are useful tools in the semantics of time-travel. Concepts like 'dancing' and 'improbability drive' also add something.
Practice Un-Attachment: This can involve such elaborate things as not doing anything noteworthy for years at a time, in order to free up time to travel back to, without sacrificing any noteworthy accomplishments. This also allows you to focus your mind on time-travel.

INTERMEDIATE QUESTS
Find a Tentative, Rational Reason to Believe You Time-Traveled: This may be harder than you think, when you really think about it. Remember, proving it to yourself is easier than proving it to someone else. The object of this quest is to prove it without a doubt for yourself.
Have a Definite Time-Travel Experience: Try traveling a short distance after any amount of effort. It requires less knowledge or technical expertise than long-distance travel, but does require a lot of authority. 'Definite' means that you know you were in the future or the noticeable past, not that it felt like anything.
Return to the Same Location at an Earlier Time: There is an odd sense of nostalgia when re-living the past.
Demonstrate Psychic Knowledge in the Past: It can be anything. You may find it is easier to focus on something unimportant.

ADVANCED QUESTS
Attend a Time-Traveler's Convention After Time-Traveling. Some of these were held in Boston in the early 2000's.

Change History for the Better: Exerting subtle influence on history can be a powerful tool. But make sure it is for the good. For example, I feel I contributed to the election of Barack Obama by saying "what about a president for the future? How about a black president?". This statement was especially powerful because I had already lived in the future.

Prove You Time-Traveled: This may involve advanced activities like pre-meditated time-travel, or carrying an object while time-traveling, or time traveling long enough to age more than ordinary, or traveling far enough back that your mother expects you to be much younger.

APPENDIX II.
DATES I'VE TRAVELED TO:
(SORTED APPROPRIATELY BY DATE OF DE-
PARTURE FOR PAST TRAVEL, AND DATE OF
ARRIVAL FOR FUTURE TRAVEL AND SEC-
ONDLY, ORDER OF TIME-TRAVEL EVENTS)

#5: The Cape Scott Case [12/2005 2nd time-
line] - Date of Cape Scott photo (1917?) -
Observations: On this date I was visited by
Jesus playing the role of a movie producer. I
was briefly on the set of some kind of movie. I
was asked if I wanted a particular costume,
and gave details. Jesus then instructed me to
hold onto a particular decorated stick for as
long as I could. When I let go, the light became more intense, and I found myself in the
awkward position of leaning onto somebody's
lap. I turned around briefly, and saw it was a
man with old-fashioned whiskers. I then
knew I had traveled significantly far back in
time. I looked around, and to my pleasure I
saw that everyone in the scene was wearing
old-fashioned clothes. We were on a hill. And
then, I noticed that my physical features had
been somewhat distorted from time-travel.
My arms and hands were somewhat more
gangly and monstrous-looking, and I could
feel my face was not exactly the saeme. In my
mind, Jessus asked me if I wanted to stay for
the photo-shoot, and I said that I did. The
camera flashed, and then I found myself once
again in the year 2005. **General Importance /
Ticket:** This scene was evidently arranged by
Jesus to provide evidence of time-travelers.
[See Photo in Main Text under Cape Scott
Case)

#7: The Car: [2013?] - 1972 - **Observations:** Clothes were faded colors. People were more introverted. People seemed to know that I had time-traveled. Dialogue: Excuse me, I don't mean to be rude. What year is this? 1972. Impressed? You know he was high! You stumpin'? You jiven'? **General Importance / Ticket:** 'Let me out of here' / Mental thought: I know the word aesthosphere. This is one of the few occasions so far where I successfully traveled quickly into the future afterwards, with the help of a friend who may have been a Djinn (I wished to time travel on this occasion, after I helped my friend get married to her fiancee).

#2: The Airplane: Sept 2007 - July 1990 (?) - **Dialogue:** What year is it? 1990. Holy fuck! What month is it? August! I -- I -- I can't say what I mean. I really thought it was September, I mean July. My younger brother looked a lot younger than me in this scene---like he used to look back when we were visiting my Dad in Washington, D.C. (In this same scene I witnessed someone fall out of the airlock of the plane without a parachute! I felt guilty because I thought it was because of the chaos of the time-stream). **General Importance / Ticket:** In both locations I was on an airplane. Apparently, I rode the waves of entropy to seek a nearly identical plane with nearly identical passengers, and then used the shockwave from arriving to boomerang myself immediately back to where I had been. I hadn't thought of this immediately, but I may have traveled forwards more years than the difference on the way back out. The name of the game was entropy.

#4: My Mother: May 25th 2009(?) -
12/28/2005 - Dialogue: What year is it?
2005. What month is it? December. Oh well,
I guess I time-traveled. Can I open Christmas
presents? You already did! And you got a car!
Just kidding. But what year is it really? It's
2005! **General Importance / Ticket:** A poem,
The ripples in the temporal pool / That mean
at first / And then forget to mean / Then
learn at last to lie or play the fool, combined
with invention of perpetual motion at an earlier endpoint and later return point (e.g. by
repeating the event, although officially the
repeated event had already happened). This
time-travel event caused me to repeat many of
the same activities, although with greater
confidence.

#1: My Own Trickery: Oct. 30, 2006 - Oct.
29, 2006 - **Observations:** On this occasion I
believed in inventing perpetual motion. This
provided a possible economic motive for the
powers that be to lose me from the current
dimension of reality. Since it was such a short
distance people doubted that I had time-
traveled, but were more likely to believe I was
doing something illegal. **General Importance / Ticket:** Invention of a powerful over-
unity device. This event became the focal
point for future time-travel occasions, which
covered more ground from a later position.

#3: The Actor: August 5th, 2009 - July 31st,
2009 - **Observations:** Seemed to occur
through someone else's decision. General
celebration and turmoil due to an actor visiting town caused me to be 'lost in the cracks.' I
had already returned once by this point. **General Importance / Ticket:** Importance of per-

petual motion for government. I conscien-
tiously used perpetual motion as a safeguard
against loss of immortality, which caused me
to gain almost a week's time at my own expense. Time travel was made easier because a
famous actor was visiting the city, causing
chaos.

#6: Unknown Date, perhaps 2009 - Feb 6,
2015 and returned. **Observations:** The trip
was an encounter with the avatar of Jesus
(who didn't look exactly like Jesus, I thought).
The trip was mysterious, as all I did was see
more advanced cellphones, and bang with my
hand on an aluminum chimney that was supposed to be on the roof of a building I would
later live in. The time travel event was later
identified as having occurred within the Feb
6th, 2016 framework of the future, as proven
by hearing the same noise from inside the
apartment. The avatar's authenticity seems to
be proven by his providing knowledge that I
would open a fan cover and rotate the fan
counter-clockwise (which I did on that same
day), and by his providing me with a differ-
ently-gendered avatar or vessel to inhabit
during my original trip. This was my only
major future travel, other than normal for-
ward-directedness. **General Importance /
Ticket:** A magical avatar assisted me.

APPENDIX III.

DIMENSIONAL TIME DIFFERENTIALS

Age / Distance to travel = Dimensional Differential

Dimensional Differential = Fraction of light speed at which one has to think /operate to successfully travel.

So, for example,

100-yr.-old traveling 100 years requires speed of light.

Same for 30 year old traveling 30 years. Etc.

100 yr-old traveling 50 years requires ½ speed of light.

Same for 30 year old traveling 15 years.

100 yr-old traveling 10 years requires 1/10 speed of light.

Same for 30 year old traveling 3 years.

100 yr-old traveling 1 year requires 1/100 speed of light.

Same for 30 year old traveling 0.3 years

A 50 year old can travel 2 years by thinking at 1/25th light speed.

Keep in mind that norms may come into play.

APPENDIX IV.

PRINCIPLES OF TIME-TRAVEL

General Inherency Principle: If time can travel, so can we!

Wave Principle: There is no big wave, just a lot of little waves, in the case in which it works.

Exceptional Rule: The properties of time-travel are only exceptional by making exceptions.

Rule of the Simplex: You can choose what seems complex about reality, once you apply a few variables to the competition

Big Wave Definition: The big wave is the sphere of influence.

Rule of Favoritism: In general, fate is on your side.

The Mastery Principle: Mastery of some kind is always present in the traveler.

Long-Term Observation: Historic events become time-travel events.

Rule of Fluxus: The rule to follow is that if the dimension has changed, it may require some travel to get back. In general, the past doesn't change unless you re-write it.

Deutch's Hypothesis: The probability of destroying a grandfather is equivalent to the probability of becoming a specific murderer and also a time-traveler.

Lloyd's Theory: Travelers may favor a world similar to the one they formerly lived in.

Criteria Problem: Those who time travel to a given location tend to resemble those who live near that dimension of reality.

Theory of Radical Explanability: That many phenomena can be explained is not coincidental, but comes down to the properties of time-travel itself: convenience is not only a tendency, it is a rule.

Rule of Aberrations: There will be no case where the energy available to measure aberrations exceeds the energy used to create those aberrations. As soon as aberrations are marked, they themselves serve as explanations.

The Rule of Inescapable Difficulty: States that travel grows increasingly rare with distance.

Corollary of Quantum Conformity: Time aberrations are permitted only when they are explainable.

Differential Un-Explainability of Aberrations: When explanations are not found, instead no explanation is available, and these aberrations tend to be able to be explained away.

General Un-Explainability: There is a limit on explanation relative to the strength of the aberration.

APPENDIX V.

GENERAL PRESCRIPTION

The future opens up. Relative to the future, you are a point in space.

If you want to time-travel in the long term, your first time travel occasion should not be fun and games.

On the other hand, if you want to have positive time-travel experiences, it might be better to begin with fun.

However, fun is not an excellent platform for formalizing mental actions, and thus may result in a degree of unpredictability.

Additional advice: wearing black is conducive, and greedy feelings are not.

Remain subtle, and practice a flexible willingness, backed up by a commitment to the basic properties of reality.

To have moral commitments just requires tacking on a little of your own extra legalese when you travel.

APPENDIX VI.

QUOTATIONS ON TIME TRAVEL

[See also quotes from my life under Language of Time Travel in the main listing].

"the fourth dimension has to do with closing the loop between non-exclusive and exclusive concepts of system"

"Themes of intelligence support, and do not detract,
from a technical use of perspective space"

"General temporal advice:
choose and you shall find"

"The Great Migration Occurs in Iterations"
---The Invention of Perpetual Motion

"Time travel is possible in any
exceptional-conditional reality
of that type: 'time' and 'travel' . "
---Nathan L. Coppedge

"The point of collapse is the synapse of the mapse [brain]"
---Figures of Agrevation

"Land is legend which has no hazard"
---Standards of Prurience

"The sun sets, near the boulevard,

where no one was discrete
[Time to go daddy
Divine words written plain
In the plain road
With irradiant truth]
What if it had been for them
As it is for me?"
---Reading the Signs with the
Conscience of the Lord

END OF APPENDIX

References, works cited...

REFERENCE BOOKS

Davies, Paul. *How to Build a Time Machine.* One of the better books, but in a conventional mold.

[Clegg, Brian. *How to Build...* Is probably a repeat of the above, but a worse edition. But it's in a better font face.]

Pickover, Clifford A. *Time: A Traveler's Guide.* Perhaps the most creative book available on time travel, other than mine.

ARTICLES CITED

http://www.quora.com/Till-now-what-evidence-points-towards-a-possibility-of-someone-having-time-travelled/answer/Neel-Bhatt-1 Source of Cape Scott Case photos

http://www.csus.edu/indiv/m/merlinos/Paradoxes%20of%20Time%20Travel.pdf Well-worth-reading, if it's still available.

INDEX

Nathan Coppedge

INDEX

> →: See only under the following
>
> **See Also:**
> At category heading, see category within Index
> At subject description, see primary contents of the book

Dummies)
health checklist, modified ($\rightarrow$Health)
qua methods ($\rightarrow$Divine Method of Time Travel)
Rhetoric of Time Travel Vis. Immortality

INCIDENTS

See Appendixes

MAGIC POWERS
Forward Travel
Hard Semantics
Horcruxes / Incarnizing
Language of Time Travel
Manipulating the Clock
Morphization
Naming the Time
Pre-Meditated Time Travel
Time Travel as a Survival Technique
Travel Patterns

MATH
Appendix (Dimensional Time Differentials)
Time Travel Energy Concepts

METAPHORS
singing ($\rightarrow$Causality and Imitation)

METHODS

Bureaucratic Slight
Divine Method of Time Travel
Five Types of Time Travel
Four Methods of Time Travel
General Archaic Method
Glitch Travel
Lemmetic Method

Manipulating the Clock
Morphization
Naming the Time
Pre-Meditated Time Travel
Primary Tractatus of Time Travel
Principal Method of Time Travel
      [84 Related Methods]
Qi / Key Method
Semantic(s) (of) Time Travel
Trinity of Methods

MISCELLANEOUS (Techniques, etc.)
Burning the Past to Feed the Future
Desirable Bargains
Forms of Repetition
Notifications
Placing Dummies
Psychological Semantics
Secret Tricks
Time Recapture Technique
Time Travel and Depression
Time Travel as a Survival Technique

MURPHY'S LAW
Time Delay

NICHES
Time Delay

PERCEPTION
Aspective Differences (art shop closes)
Doing the Wave
Length of Time
Matums
Panoptic Space
Unitation (time is visible)

Dimensional Time-Travel Toolkit

PHENOMENA
>Aspective Differences
>Slim Picketings (light on the horizon, etc.)
>States of Awareness
>Travel Patterns

PHILOSOPHICAL TOOLS
>Power Majore Vs. Authority
>>Minusculi

PRECAUTIONS
>Notifications
>Placing Dummies

PROBLEMS
>Criteria Problem and
>>Inescapable Difficulties
>Darwinistic Problem
>Foreshortening or
>>Trapdoor Problem
>Slide Travel
>Time Traps
>Time Trouble

PSYCHOLOGY
>Fractionistic Origin of Time Travel
>Psychological Factors
>States of Awareness

SECRETS
>Dirty Secrets to Time
>>Travel
>Power Majore Vs. Authority
>>Minusculi
>(The) Secret of November
>Secret Tricks
>"Shaving"

Dimensional Time-Travel Toolkit

END OF INDEX

BIO

Nathan Coppedge is a philosopher, artist, inventor, and poet in some capacity. He may be remembered for inventing perpetual motion. He lives in New Haven. In this book he recounts his methods for time-travel, as well as notable events including the Cape Scott Case in which he was photographed in 1917, and a time he attended a "Time-Traveler's Convention" in Boston and a time he was shown the future by a Christian avatar.

www.ingramcontent.com/pod-product-compliance
Lightning Source LLC
Chambersburg PA
CBHW072303200526
45168CB00014B/338